I0488144

Status and Understanding of Groundwater Quality in the San Diego Drainages Hydrogeologic Province, 2004: California GAMA Priority Basin Project

By Michael T. Wright and Kenneth Belitz

A product of the California Groundwater Ambient Monitoring and Assessment (GAMA) Program

Prepared in cooperation with the California State Water Resources Control Board

Scientific Investigations Report 2011–5154

U.S. Department of the Interior
U.S. Geological Survey

U.S. Department of the Interior
KEN SALAZAR, Secretary

U.S. Geological Survey
Marcia K. McNutt, Director

U.S. Geological Survey, Reston, Virginia: 2011

For more information on the USGS—the Federal source for science about the Earth, its natural and living resources, natural hazards, and the environment, visit http://www.usgs.gov or call 1–888–ASK–USGS.

For an overview of USGS information products, including maps, imagery, and publications, visit http://www.usgs.gov/pubprod

To order this and other USGS information products, visit http://store.usgs.gov

Suggested citation:
Wright, M.T., and Belitz, K., 2011, Status and understanding of groundwater quality in the San Diego Drainages Hydrogeologic Province, 2004—California GAMA Priority Basin Project: U.S. Geological Survey Scientific Investigations Report 2011–5154, 100 p.

Contents

Contents—Continued

Figures

Figures—Continued

Figures—Continued

Tables

Conversion Factors, Datums, and Abbreviations and Acronyms

Conversion Factors

Inch/Pound to SI

Multiply	By	To obtain
Length		
inch (in.)	2.54	centimeter (cm)
inch (in.)	25.4	millimeter (mm)
foot (ft)	0.3048	meter (m)
mile (mi)	1.609	kilometer (km)
Area		
square mile (mi^2)	2.590	square kilometer (km^2)
Flow rate		
gallon per minute (gal/min)	0.06309	liter per second (L/s)

SI to Inch/Pound

Multiply	By	To obtain
Length		
millimeter (mm)	0.03937	inch (in.)
kilometer (km)	0.6214	mile (mi)
meter (m)	3.281	foot (ft)
Volume		
liter (L)	0.2642	gallon (gal)
Mass		
kilogram (kg)	2.205	pound avoirdupois (lb)

Temperature in degrees Celsius (°C) may be converted to degrees Fahrenheit (°F) as follows:

$$°F=(1.8×°C)+32.$$

Specific conductance is given in microsiemens per centimeter at 25 degrees Celsius (µS/cm at 25°C).

Concentrations of chemical constituents in water are given either in milligrams per liter (mg/L) or micrograms per liter (µg/L). Milligrams per liter is equivalent to parts per million (ppm), and micrograms per liter is equivalent to parts per billion (ppb).

cm^3 STP g^{-1}	cubic centimeters at standard temperature and pressure per gram
δiE	delta notation; the ratio of the heavier isotope (i) to the more common lighter isotope of an element (E), relative to a standard reference material, expressed as per mil
mL	milliliter
pCi/L	picocuries per liter
per mil	parts per thousand
pmc	percent modern carbon
TU	tritium unit
%	percent

Conversion Factors, Datums, and Abbreviations and Acronyms—Continued

Datums

Vertical coordinate information is referenced to the North American Vertical Datum of 1988 (NAVD 88).

Horizontal coordinate information is referenced to the North American Datum of 1983 (NAD 83).

Abbreviations and Acronyms

AL-US	U.S. Environmental Protection Agency action level
BLS	below land surface
DG	CDPH data from well sampled by GAMA
DPH	CDPH data from well not sampled by GAMA
E	estimated or having a higher degree of uncertainty
GAMA	Groundwater Ambient Monitoring and Assessment Program
HAL-US	U.S. Environmental Protection Agency lifetime health advisory level
HBSL	health-based screening level
LRL	laboratory reporting level
LSD	land-surface datum
LT-MDL	long-term method detection level
MCL-CA	California Department of Public Health maximum contaminant level
MCL-US	U.S. Environmental Protection Agency maximum contaminant level
MDL	method detection limit
MRL	minimum reporting level
NL-CA	California Department of Public Health notification level
NWIS	National Water Information System (USGS)
PSW	public-supply wells
RPD	relative percentage difference
RSD	relative standard deviation
RSD5-US	U.S. Environmental Protection Agency risk-specific dose at a risk factor of 10^{-5}
SDALLV	Alluvial Basins study area
SDALLVU	Alluvial Basins study area understanding well
SDHDRK	Hard Rock study area
SDHDRKU	Hard Rock study area understanding well
SDTEM	Temecula Valley study area
SDTEMFP	Temecula Valley study area flow path well
SDWARN	Warner Valley study area
SMCL-CA	California Department of Public Health secondary maximum contaminant level
SMCL-US	U.S. Environmental Protection Agency secondary maximum contaminant level
US	United States
>	greater than
≥	greater than or equal to
<	less than
≤	less than or equal to

Conversion Factors, Datums, and Abbreviations and Acronyms—Continued

Organizations

CDPH	California Department of Public Health (was California Department of Health Services prior to July 1, 2007)
CDWR	California Department of Water Resources
LLNL	Lawrence Livermore National Laboratory
NAWQA	National Water Quality Assessment (USGS)
NWQL	National Water Quality Laboratory (USGS)
SWRCB	State Water Resources Control Board (California)
USEPA	U.S. Environmental Protection Agency
USGS	U. S. Geological Survey

Selected Chemical Names

DO	dissolved oxygen
Fe	iron
Nitrate-N	nitrate as nitrogen
Nitrite-N	nitrite as nitrogen
Mn	manganese
MTBE	methyl *tert*-butyl ether
PCE	perchloroethene (tetrachloroethene)
TCE	trichloroethene
TDS	total dissolved solids
THM	trihalomethane
VOC	volatile organic compound

Status and Understanding of Groundwater Quality in the San Diego Drainages Hydrogeologic Province, 2004: California GAMA Priority Basin Project

By Michael T. Wright and Kenneth Belitz

Abstract

Groundwater quality in the approximately 3,900-square-mile (mi^2) San Diego Drainages Hydrogeologic Province (hereinafter San Diego) study unit was investigated from May through July 2004 as part of the Priority Basin Project of the Groundwater Ambient Monitoring and Assessment (GAMA) Program. The study unit is located in southwestern California in the counties of San Diego, Riverside, and Orange. The GAMA Priority Basin Project is being conducted by the California State Water Resources Control Board in collaboration with the U.S. Geological Survey (USGS) and the Lawrence Livermore National Laboratory.

The GAMA San Diego study was designed to provide a statistically robust assessment of untreated-groundwater quality within the primary aquifer systems. The assessment is based on water-quality and ancillary data collected by the USGS from 58 wells in 2004 and water-quality data from the California Department of Public Health (CDPH) database. The primary aquifer systems (hereinafter referred to as the primary aquifers) were defined by the depth interval of the wells listed in the California Department of Public Health (CDPH) database for the San Diego study unit. The San Diego study unit consisted of four study areas: Temecula Valley (140 mi^2), Warner Valley (34 mi^2), Alluvial Basins (166 mi^2), and Hard Rock (850 mi^2). The quality of groundwater in shallow or deep water-bearing zones may differ from that in the primary aquifers. For example, shallow groundwater may be more vulnerable to surficial contamination than groundwater in deep water-bearing zones.

This study had two components: the *status assessment* and the *understanding assessment*. The first component of this study—the *status assessment* of the current quality of the groundwater resource—was assessed by using data from samples analyzed for volatile organic compounds (VOC), pesticides, and naturally occurring inorganic constituents, such as major ions and trace elements. The status assessment is intended to characterize the quality of groundwater resources within the primary aquifers of the San Diego study unit, not the treated drinking water delivered to consumers by water purveyors. The second component of this study—the *understanding assessment*—identified the natural and human factors that affect groundwater quality by evaluating land use, well construction, and geochemical conditions of the aquifer. Results from these evaluations were used to help explain the occurrence and distribution of selected constituents in the study unit.

Relative-concentrations (sample concentration divided by benchmark concentration) were used as the primary metric for relating concentrations of constituents in groundwater samples to water-quality benchmarks for those constituents that have Federal and (or) California benchmarks. For organic and special-interest constituents, relative-concentrations were classified as high (> 1.0), moderate (> 0.1 and ≤ 1.0), and low (≤ 0.1). For inorganic constituents, relative concentrations were classified as high (> 1.0), moderate (> 0.5 and ≤ 1.0), and low (≤ 0.5). Grid-based and spatially weighted approaches were then used to evaluate the proportion of the primary aquifers (aquifer-scale proportions) with high, moderate, and low relative-concentrations for individual compounds and classes of constituents.

One or more of the inorganic constituents with health-based benchmarks were high (relative to those benchmarks) in 17.6 percent of the primary aquifers in the Temecula Valley, Warner Valley, and Alluvial Basins study areas (hereinafter also collectively referred to as the Alluvial Fill study areas because they are composed of alluvial fill aquifers), and in 25.0 percent of the Hard Rock study area. Inorganic constituents with health-based benchmarks that were frequently detected at high relative-concentrations included vanadium (V), arsenic (As), and boron (B). Vanadium and As concentrations were not significantly correlated to either urban or agricultural land use indicating natural sources as the primary contributors of these constituents to groundwater. The positive correlation of B concentration to urban land-use was significant which indicates that anthropogenic activities are a contributing source of B to groundwater. The correlation of V, As and B concentrations to pH was positive, indicating that in alkaline groundwater these constituents are being desorbed from, or being inhibited from adsorbing to, particle surfaces.

Inorganic constituents with aesthetic benchmarks that were detected at high relative-concentrations include manganese (Mn), iron (Fe), and total dissolved solids (TDS). In the Alluvial Fill study areas, Mn and TDS were detected at high relative-concentrations in 13.7 percent of the primary aquifers, and Fe in 6.9 percent of the primary aquifers. In the Hard Rock study area, Mn was detected at high relative-concentrations in 33.3 percent of the primary aquifers, and TDS in 16.7 percent; Fe was not detected at high relative-concentrations. Total dissolved solids concentrations were significantly correlated to agricultural land use suggesting that agricultural practices are a contributing source of TDS to groundwater. Manganese and Fe concentrations were highest in groundwater with low dissolved oxygen and pH indicating that the reductive dissolution of oxyhydroxides may be an important mechanism for the mobilization of Mn and Fe in groundwater. TDS concentrations were highest in shallow wells and in modern (< 50 yrs) groundwater which indicates anthropogenic activities as a source of TDS concentrations in groundwater.

The relative-concentrations of organic constituents with health-based benchmarks were high in 3.0 percent of the primary aquifers in the Alluvial Fill study areas. A single detection in the Alluvial Basins study area of the discontinued gasoline oxygenate methyl *tert*-butyl ether (MTBE) was the only organic constituent detected at a high relative-concentration; high relative-concentrations of these constituents were not detected in the Hard Rock study area. Twelve of 88 VOCs and 14 of 123 pesticides and pesticide degradates analyzed in grid wells were detected. Chloroform was the only VOC detected in more than 10 percent of the grid wells. The herbicides simazine, atrazine, and prometon were each detected in greater than 10 percent of the grid wells. Perchlorate was detected in 22 percent of the grid wells sampled.

The *understanding assessment* showed a significant correlation of trihalomethanes (THMs) and solvents to urban land-use, indicating that detections of these constituents are more likely to occur in groundwater underlying urbanized areas of the study unit. MTBE concentrations were negatively correlated to the distance from the nearest leaking underground fuel tank, indicating that point sources are the most significant contributing factor for MTBE concentrations to groundwater in the study unit. The positive correlation of THM and herbicide concentrations to modern groundwater was significant, as was the negative correlation of herbicide concentrations to pH and anoxic groundwater. The negative correlation of herbicides to pH and anoxic groundwater was likely due to the fact that these constituents were detected more frequently in shallow wells where groundwater conditions tend to be oxic with relatively low pH.

Introduction

To assess the quality of ambient groundwater in aquifers used for drinking-water supply and to establish a baseline groundwater-quality monitoring program, the State Water Resources Control Board (SWRCB), in collaboration with the U.S. Geological Survey (USGS) and Lawrence Livermore National Laboratory (LLNL), implemented the Groundwater Ambient Monitoring and Assessment (GAMA) Program (State of California, 2011, at http://www.waterboards.ca.gov/gama). The statewide GAMA program currently consists of three projects: the GAMA Priority Basin Project, conducted by the USGS (U.S. Geological Survey, 2011, at http://ca.water.usgs. gov/gama/); the GAMA Domestic Well Project, conducted by the SWRCB; and GAMA Special Studies, conducted by LLNL. Statewide, the Priority Basin Project primarily focused on the deep part of the groundwater resource, and the SWRCB Domestic Well Project generally focused on the shallow aquifer systems. Shallow groundwater wells, such as private domestic and environmental monitoring wells, may be particularly at risk because of surficial contamination. As a result, concentrations of contaminants, such as VOCs and nitrate, in shallow wells can be higher than in wells screened in the deep primary aquifers (Landon and others, 2010).

The SWRCB initiated the GAMA Program in 2000 in response to a legislative mandate (State of California, 1999, 2001a, Supplemental Report of the 1999 Budget Act 1999–00 Fiscal Year). The GAMA Priority Basin Project was initiated in response to the Groundwater Quality Monitoring Act of 2001 (State of California, 2001b, Sections 10780–10782.3 of the California Water Code, Assembly Bill 599) to assess and monitor the quality of groundwater in California. The GAMA Priority Basin Project is a comprehensive assessment of statewide groundwater quality designed to improve understanding and identification of risks to groundwater resources and to increase the availability of information about groundwater quality to the public. For the Priority Basin Project, the USGS, in collaboration with the SWRCB, developed the monitoring plan to assess groundwater basins through direct and other statistically reliable sampling approaches (Belitz and others, 2003; State Water Resources Control Board, 2003). Additional partners in the GAMA Priority Basin Project include the California Department of Public Health (CDPH), the California Department of Pesticide Regulation (CDPR), the California Department of Water Resources (CDWR), and local water agencies and well owners (Kulongoski and Belitz, 2004).

The range of hydrologic, geologic, and climatic conditions in California must be considered in an assessment of groundwater quality. Belitz and others (2003) partitioned the State into ten hydrogeologic provinces, each with distinctive hydrologic, geologic, and climatic characteristics (fig. 1).

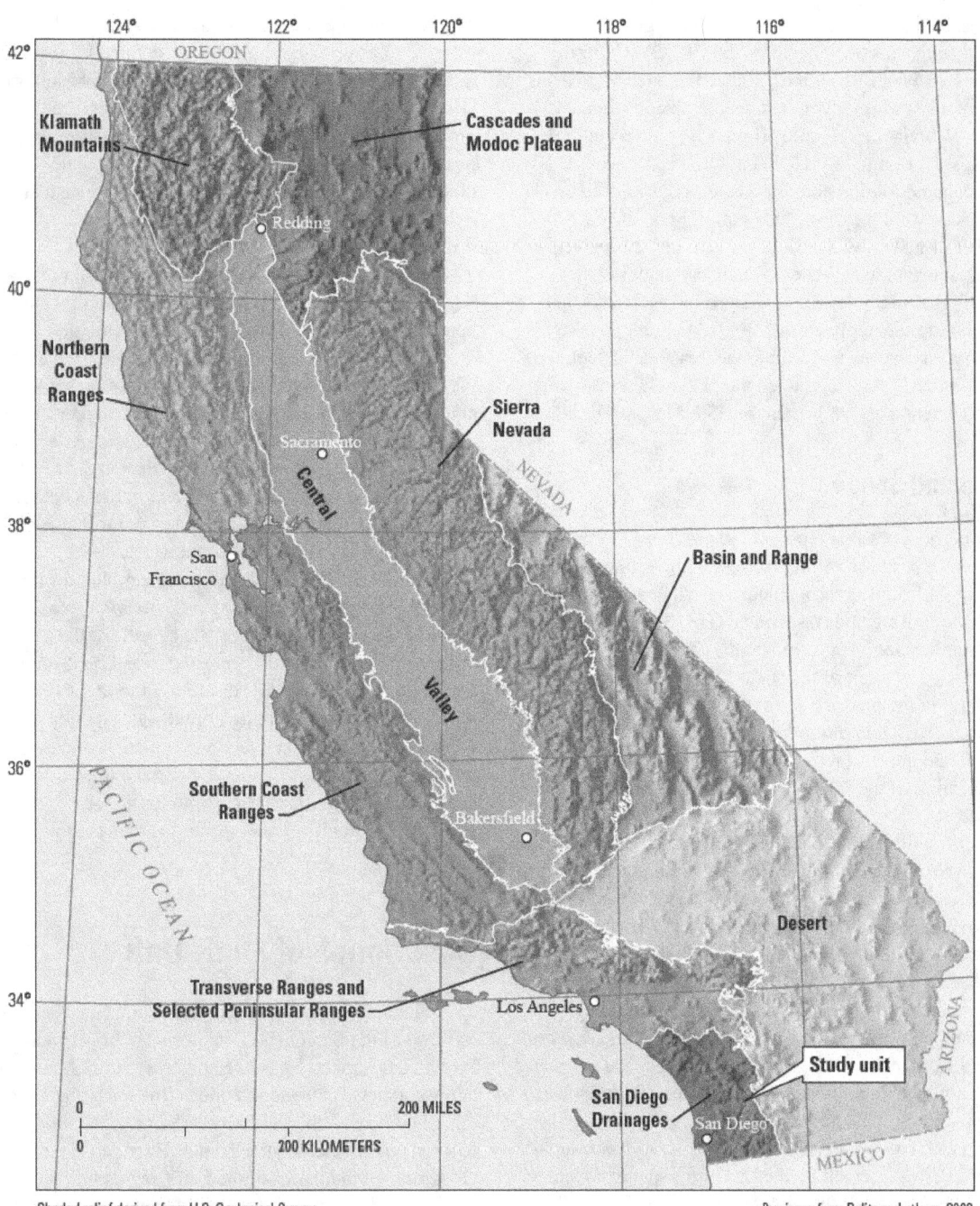

Figure 1. Location of the California hydrogeologic provinces and the San Diego Groundwater Ambient Monitoring and Assessment (GAMA) study unit, California.

All of these hydrogeologic provinces contain groundwater basins and subbasins designated by the CDWR (California Department of Water Resources, 2003). Groundwater basins generally consist of relatively permeable, unconsolidated deposits of alluvial or volcanic origin. Eighty percent of California's approximately 16,000 public-supply wells (PSW) are in designated groundwater basins. Groundwater basins and subbasins were prioritized for sampling on the basis of the number of PSWs, with secondary consideration given to municipal groundwater use, agricultural pumping, the number of historically leaking underground fuel tanks, and registered pesticide applications (Belitz, and others, 2003). The 116 priority basins and additional areas outside defined groundwater basins were grouped into 35 study units, which include approximately 95 percent of PSWs in California.

Purpose and Scope

The purposes of this report are to provide a (1) study unit description: description of the hydrogeologic setting of the San Diego Drainages Hydrogeologic Province Groundwater Ambient Monitoring and Assessment (GAMA) study unit (hereinafter San Diego study unit) (fig. 1), (2) *status assessment*: assessment of the status of the current quality of groundwater in the primary aquifer systems in the San Diego study unit, and (3) *understanding assessment*: identification of the natural and human factors affecting groundwater quality and explanation of the relations between water quality and those factors.

Water-quality data for samples collected by the USGS for the GAMA Program in the San Diego study unit and details of sample collection, analysis, and quality-assurance procedures for the San Diego study unit are reported by Wright and others (2005). Utilizing those same data, this report describes methods used in designing the sampling network, identifying CDPH data for use in the *status assessment*, estimating aquifer-scale proportions of relative-concentrations, analyzing ancillary datasets, classifying groundwater age, and assessing the status and understanding of groundwater quality by using statistical and graphical approachess.

The *status assessment* includes analyses of water-quality data for 47 PSWs selected by the USGS for spatial coverage of one well per grid cell (hereinafter referred to as USGS-grid wells) across the San Diego study unit. Samples were collected for analysis of anthropogenic constituents, such as volatile organic compounds (VOC) and pesticides, and naturally occurring inorganic constituents such as major ions and trace elements. Water-quality data from 23 PSWs in the California Department of Public Health (CDPH) database also were used to supplement data collected by USGS for the GAMA program. The resulting set of water-quality data from USGS-grid wells and selected CDPH wells was considered to be representative of the primary aquifer systems in the San

Diego study unit; the primary aquifer systems (hereinafter referred to as primary aquifers) are defined by the depth interval of the wells listed in the CDPH database for the San Diego study unit. GAMA *status assessments* are designed to provide a statistically robust characterization of groundwater quality in the primary aquifers at the basin-scale (Belitz and others, 2003). The statistically robust design also allows basins to be compared and results to be synthesized at regional and statewide scales.

To provide context, the water-quality data discussed in this report were compared to State and Federal drinking-water benchmarks, both regulatory and non-regulatory, for treated drinking water. The assessments in this report characterize the quality of untreated groundwater resources in the primary aquifers within the study unit, not the treated drinking water delivered to consumers by water purveyors. After withdrawal from the ground, water typically is treated, disinfected, and (or) blended with other waters to maintain acceptable water quality. Benchmarks apply to treated water that is served to the consumer, not to untreated groundwater.

In addition to the 47 grid-wells sampled for the *status assessment*, the *understanding assessment* also uses data from the 11 wells sampled by the USGS for the purposes of understanding (hereinafter referred to as USGS-understanding wells). Data from these wells are used to identify the natural and human factors affecting groundwater quality and to explain the relations between water quality and selected potential explanatory factors. Potential explanatory factors examined included land use, depth to the top of the uppermost open interval, indicators of groundwater age, and geochemical conditions.

Description of Study Unit

The San Diego study unit boundaries are the same as those of the San Diego Drainages Hydrogeologic Province described by Belitz and others (2003) and covers approximately 3,900 square miles (mi^2). The San Diego study unit encompasses the majority of San Diego County, as well as parts of southwestern Orange and Riverside Counties (fig. 2). Geographic boundaries of the San Diego study unit are the Transverse Ranges and Selected Peninsular Ranges Province to the north, the Desert Province to the east, the country of Mexico to the south, and the Pacific Ocean to the west.

The climate in the coastal areas of the San Diego study unit generally is mild, with temperatures averaging 64 degrees Fahrenheit (°F) and average annual precipitation ranging from 10 to 13 inches (in.) (California Regional Water Quality Control Board, San Diego Region, 1994). In the eastern part of the study unit, annual temperatures in the Peninsular Ranges average 55 °F, with average annual precipitation of approximately 45 in.

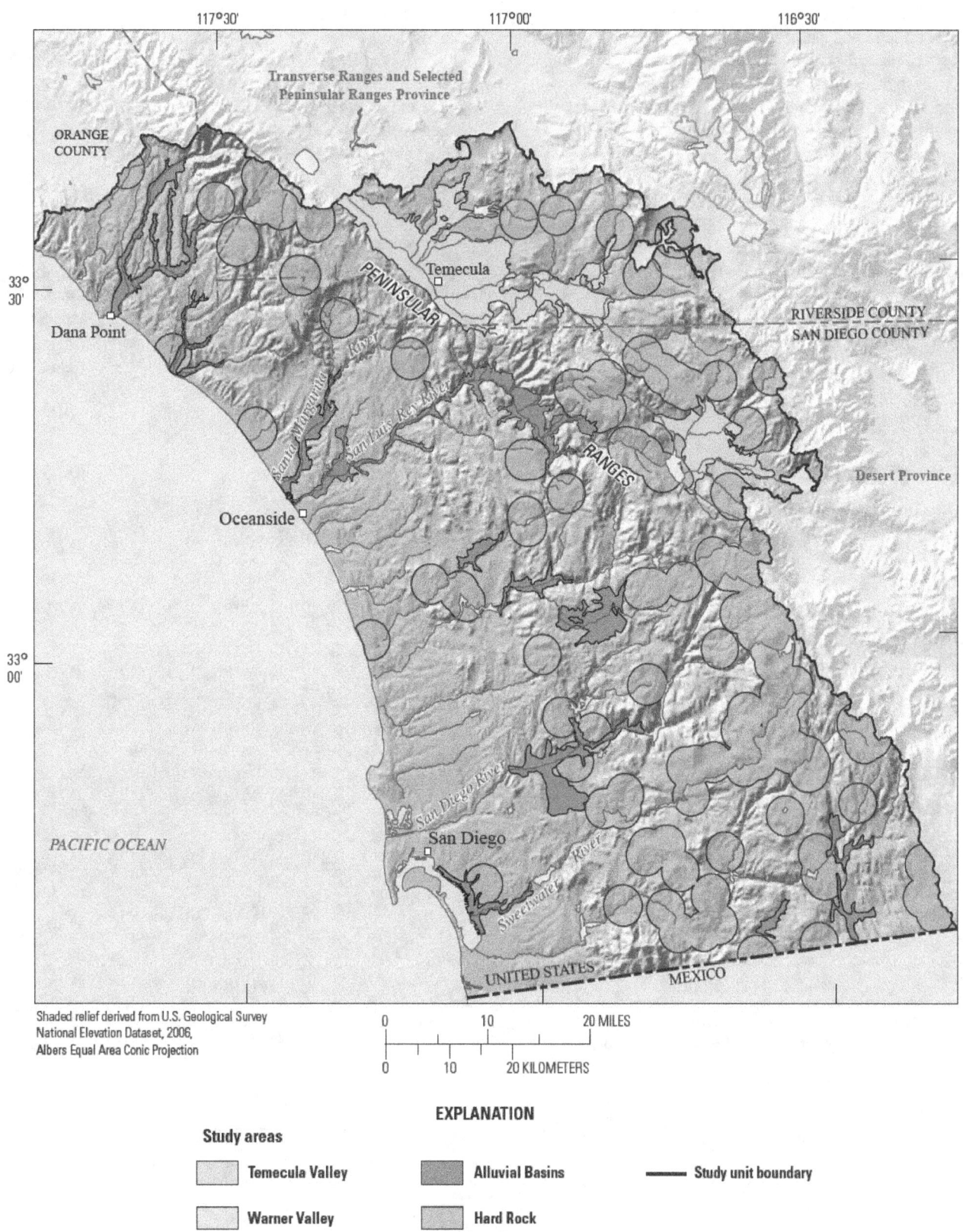

Figure 2. Geographic features and study area boundaries of the San Diego Groundwater Ambient Monitoring and Assessment (GAMA) study unit, California.

The San Diego study unit is drained by a number of creeks and rivers, including the Santa Margarita and San Luis Rey Rivers in the north, and the San Diego and Sweetwater Rivers in the south (fig. 2). Runoff in the study unit is attributed mainly to rainfall; however, smaller amounts of runoff come from urban water use, snowmelt, and artesian springs. Groundwater and surface-water flow direction is primarily from the mountainous east towards the west and the Pacific Ocean. Groundwater recharges in the study unit by precipitation, irrigation returns, infiltration of reservoir and river water, and engineered recharge. Groundwater primarily discharges through pumping from wells.

The San Diego study unit is composed of relatively small groundwater basins that underlie approximately 400 mi^2 of land surface, corresponding to the Temecula Valley, Warner Valley, and Alluvial Basins study areas (fig. 2). In addition, a part of the groundwater resources in the San Diego study unit are in areas outside of defined groundwater basins. This area underlies approximately 850 mi^2 of the study unit land surface and was defined as all areas located outside a CDWR-defined groundwater basin, but within 1.9 miles (mi) of a PSW documented in the CDPH database, and corresponds to the Hard Rock study area (fig. 2).

The land use in the study unit is 7 percent agricultural, 84 percent natural, and 9 percent urban based on classification by the USGS National Land Cover Data (Vogelman and others, 2001; Price and others, 2003) (fig. 3A). The natural land-use areas are mostly shrub land, with lesser amounts of evergreen forest and grass lands. Natural land-use is most predominant in the eastern parts of the study unit (fig. 4A–C).

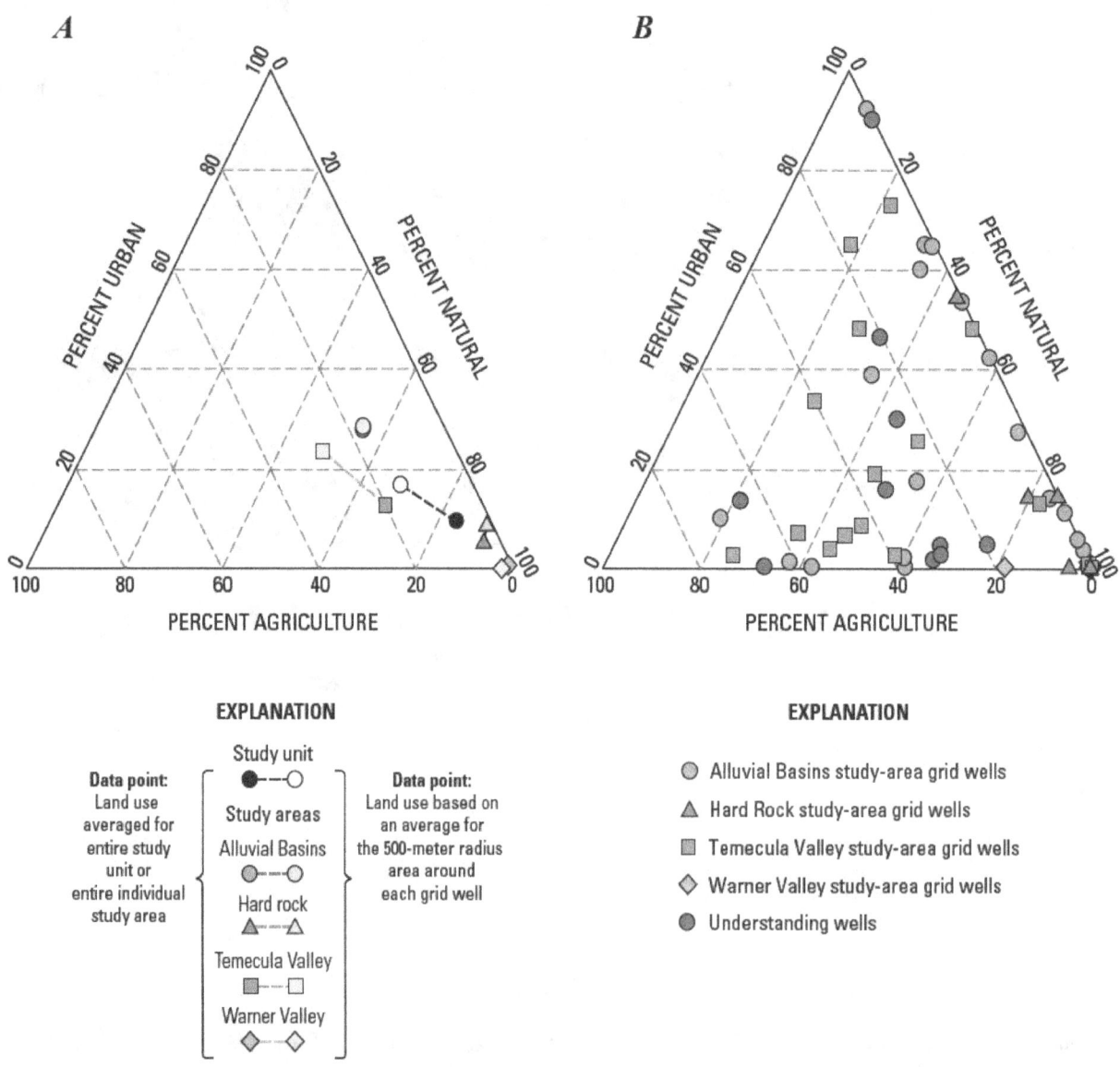

Figure 3A–B. Ternary diagram of proportions of urban, agricultural, and natural land-uses in the San Diego Groundwater Ambient Monitoring and Assessment (GAMA) study unit, California. (A) Study unit and study areas, (B) wells.

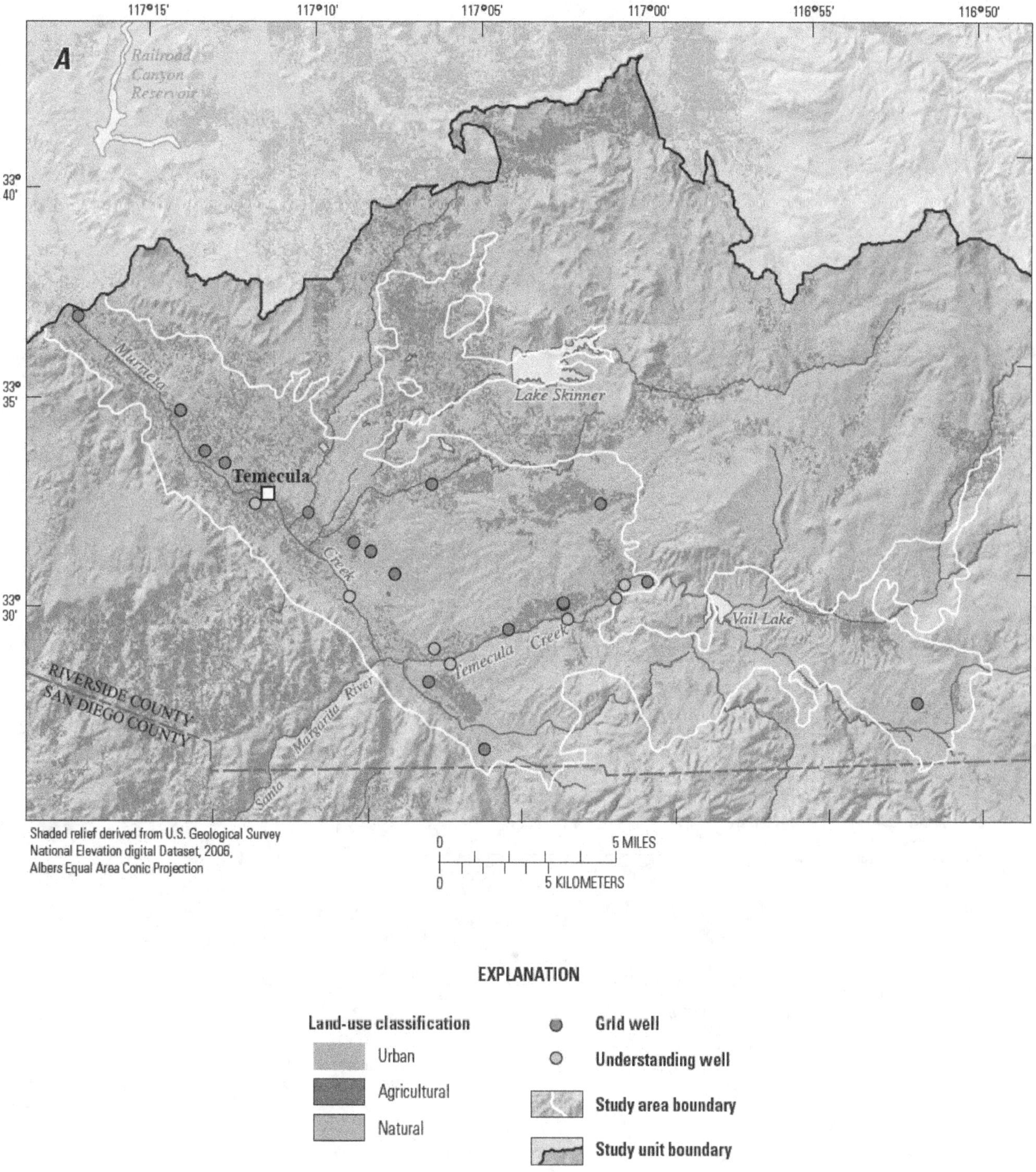

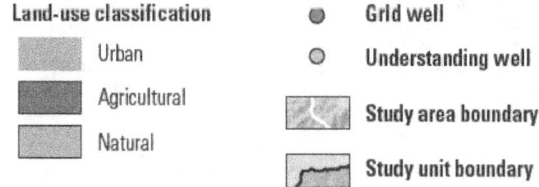

Figure 4A–C. Land-use classification in the San Diego Groundwater Ambient Monitoring and Assessment (GAMA) study unit, and locations of grid and understanding wells. (A) Temecula Valley, (B) Warner Valley, (C) Alluvial Basins, and Hard Rock study areas.

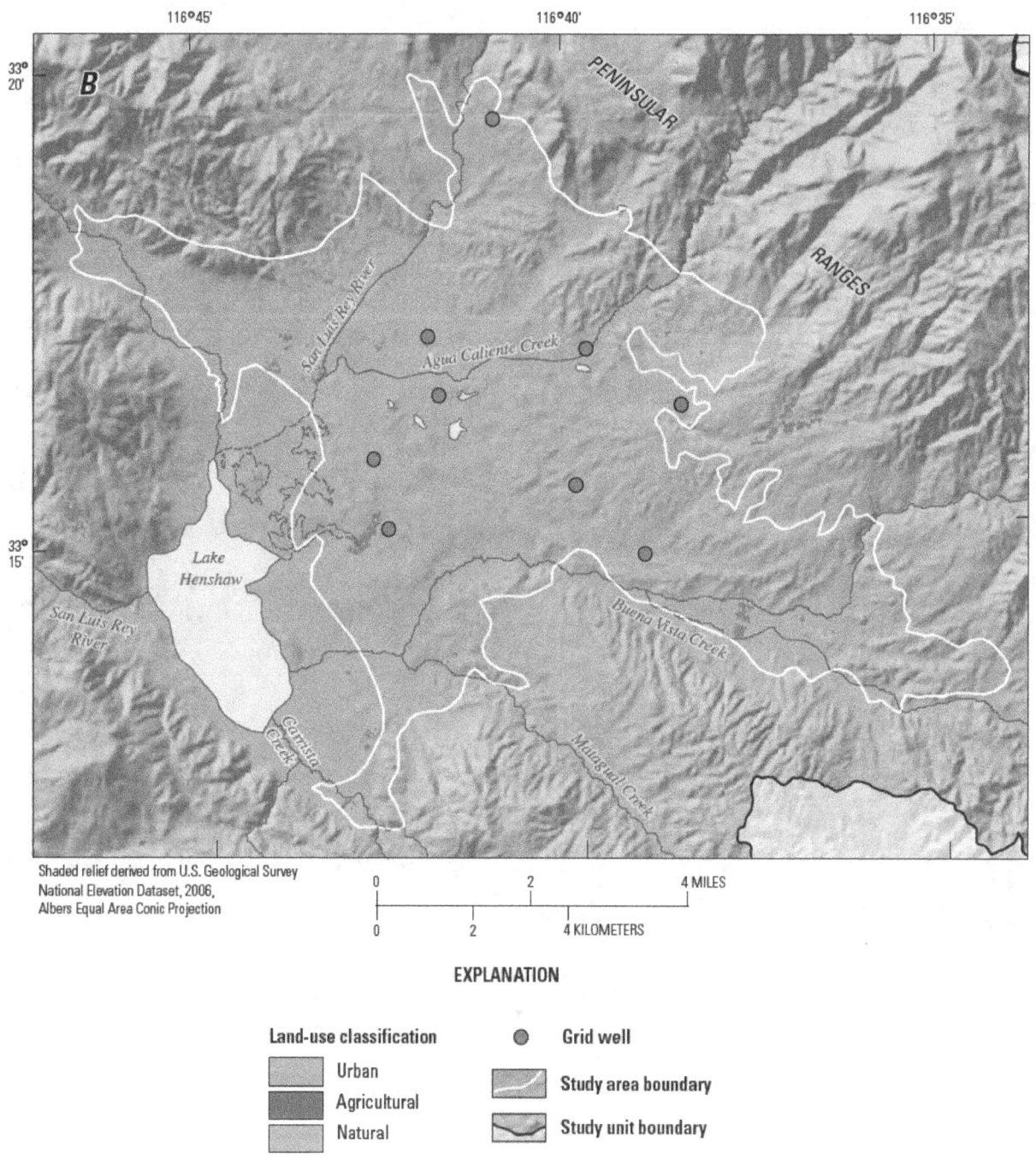

Shaded relief derived from U.S. Geological Survey
National Elevation Dataset, 2006,
Albers Equal Area Conic Projection

EXPLANATION

Land-use classification   ● Grid well

    Urban   Study area boundary

    Agricultural

    Natural   Study unit boundary

Figure 4*A–C.*—Continued

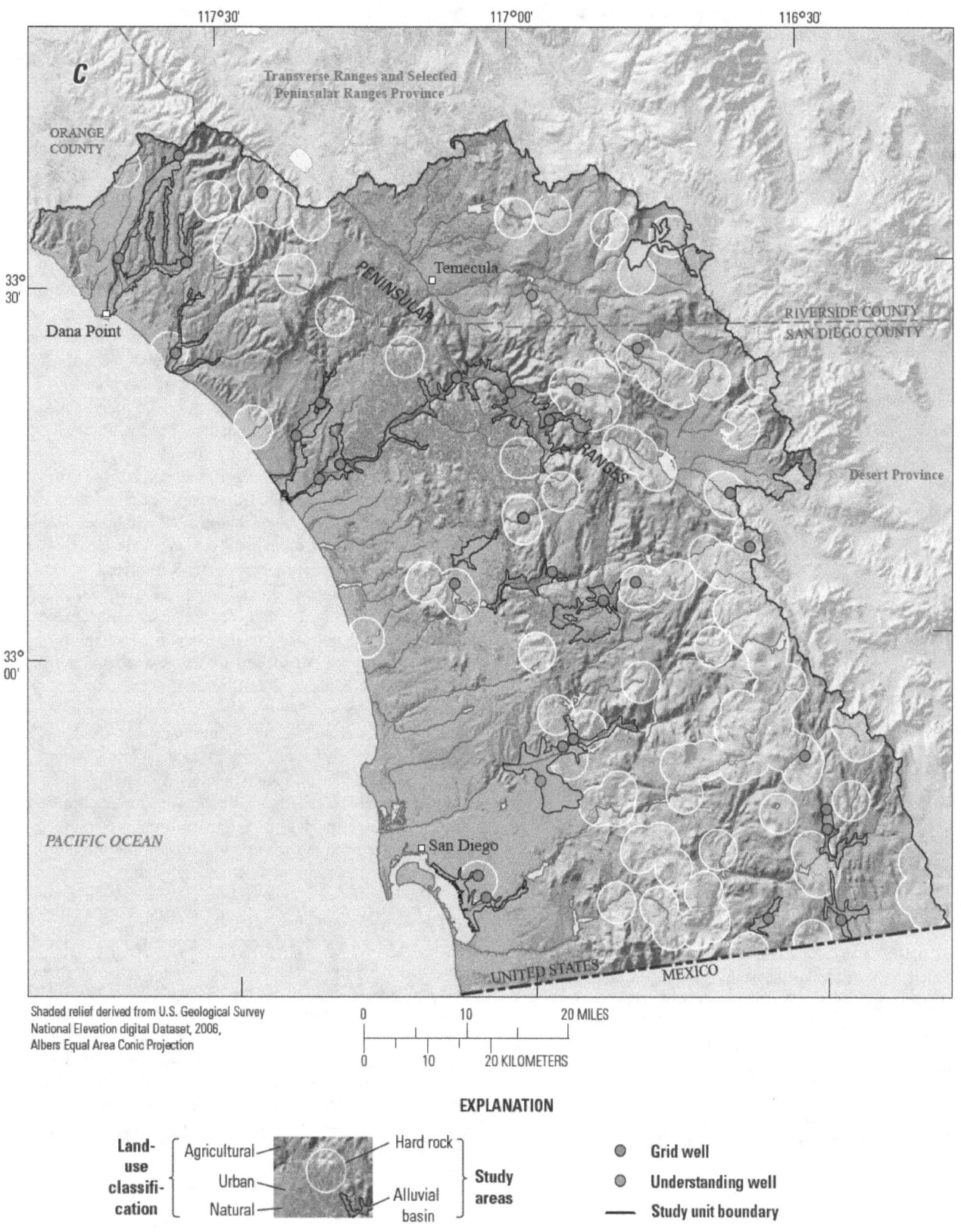

Shaded relief derived from U.S. Geological Survey
National Elevation digital Dataset, 2006,
Albers Equal Area Conic Projection

EXPLANATION

Figure 4*A–C*.—Continued

Agricultural land-use in the study unit is equal parts orchards and pasture land, with a small percentage of row crops. Urban land-use primarily is found in the coastal areas of the study unit and the largest urban center is the San Diego metropolitan area. The majority of land use in all study areas is natural; the Warner Valley and Hard Rock study areas are classified as 99 and 91 percent natural, respectively (fig. 3A). The Alluvial Basins and Temecula Valley study areas are the most urbanized in the San Diego study unit (28 and 13 percent, respectively); the largest amount of agricultural land-use also is in these study areas (17 and 20 percent respectively).

Description of Study Areas

The boundaries of the Temecula Valley study area (fig. 2) are the same as those of the Temecula Valley groundwater basin as described by the California Department of Water Resources, (2004a). The Temecula Valley study area primarily is in southwestern Riverside County with a very small part of the basin extending into northern San Diego County. The Temecula Valley study area covers approximately 140 mi^2 and is bounded by the relatively impermeable rocks of the Peninsular Ranges on three sides. The main water-bearing units are Quaternary alluvium that is estimated to be as great as 2,500 feet (ft) thick; generally it is unconfined except in areas where faults cut across the basin (California Department of Water Resources, 1956; Kennedy, 1977). Rock types that bound the groundwater-bearing deposits in the study area include Mesozoic granites and gabbros and Jurassic marine sedimentary rocks (fig. 5A) (Saucedo, 2000). Sources of groundwater recharge in the basin include percolation of precipitation, infiltration of irrigation and domestic return water, and engineered recharge from spreading basins along Temecula Creek. Groundwater primarily discharges through groundwater pumping. Average annual precipitation ranges from 7 to 15 in. Surface water drains to several creeks, including Temecula and Murrieta Creeks that discharge into the Santa Margarita River, which then flows westward out of the valley.

The boundaries of the Warner Valley study area (fig. 2) are the same as those of the Warner Valley groundwater basin, which is located in northeastern San Diego County (California Department of Water Resources, 2004b). The Warner Valley study area has a surface area of 37 mi^2; it is bounded on the west by Lake Henshaw and on all other sides by the crystalline rocks of the Peninsular Ranges. The main water-bearing unit consists of alluvium and residuum (California Department of Water Resources, 1971). The alluvium is as great as 900 ft thick and generally is unconsolidated. The crystalline rocks

that bound the groundwater-bearing deposits in this study area consist primarily of Mesozoic granite and metamorphic rocks of pre-Cenozoic age (fig. 5B) (Saucedo, 2000). Sources of groundwater recharge include percolation of precipitation, and river and stream runoff. Groundwater discharges primarily through groundwater pumping. Annual precipitation ranges from 15 to 21 in. The Warner Valley study area is primarily drained by the Agua Caliente and Buena Vista Creeks, and the San Luis Rey River, all of which flow westward into Lake Henshaw.

The Alluvial Basins study area (fig. 2) is composed of all alluvial basins in the study unit that have one or more PSWs. The 12 groundwater basins in this study area are the San Juan, San Mateo, Santa Margarita, San Luis Rey, San Pasqual, Santa Maria, San Diego River, El Cajon, Sweetwater, Cottonwood, Campo, and Potrero Valleys (California Department of Water Resources, 2003). The collective surface area of the study area is approximately 166 mi^2, with individual basins ranging in area from as small as 3 mi^2 (California Department of Water Resources, 2004c), to as large as 46 mi^2 (California Department of Water Resources, 2004d). The main water-bearing units are Quaternary age alluvium and residuum, with an average thickness of alluvium that ranges from approximately 15 ft in the San Mateo Valley groundwater basin (California Department of Water Resources, 1991) to 60 ft in the San Luis Rey groundwater basin (Izbicki, 1985). Inland alluvial basins generally are bound by the Mesozoic granites of the Peninsular Ranges, whereas coastal alluvial basins generally are bounded by Cenozoic-aged sedimentary rocks (fig. 5C) (Saucedo, 2000). Sources of groundwater recharge include percolation of precipitation, river and stream runoff, agricultural and domestic returns, discharge of wastewater to rivers, and septic systems. Groundwater primarily discharges through groundwater pumping. The average annual precipitation in these basins range from as little as 8 in. to as great as 21 in. Runoff from precipitation primarily is drained to the southwest towards the Pacific Ocean, but some basins are internally drained.

The Hard Rock study area (fig. 2) consists of all areas outside of CDWR-defined groundwater basins that are within 3 km of a PSW. The study area covers approximately 850 mi^2 and most of the study area is in the inland areas of the study unit. Surficial geology in the study area primarily is composed of granitic and metamorphic rocks with small amounts of Mesozoic volcanic and Cenozoic marine sedimentary rocks (fig. 5C). Well completion reports for the PSWs sampled by the GAMA program indicate that wells are withdrawing water primarily from fractured and decomposed granite. Sources of groundwater recharge include percolation of precipitation, and river and stream runoff.

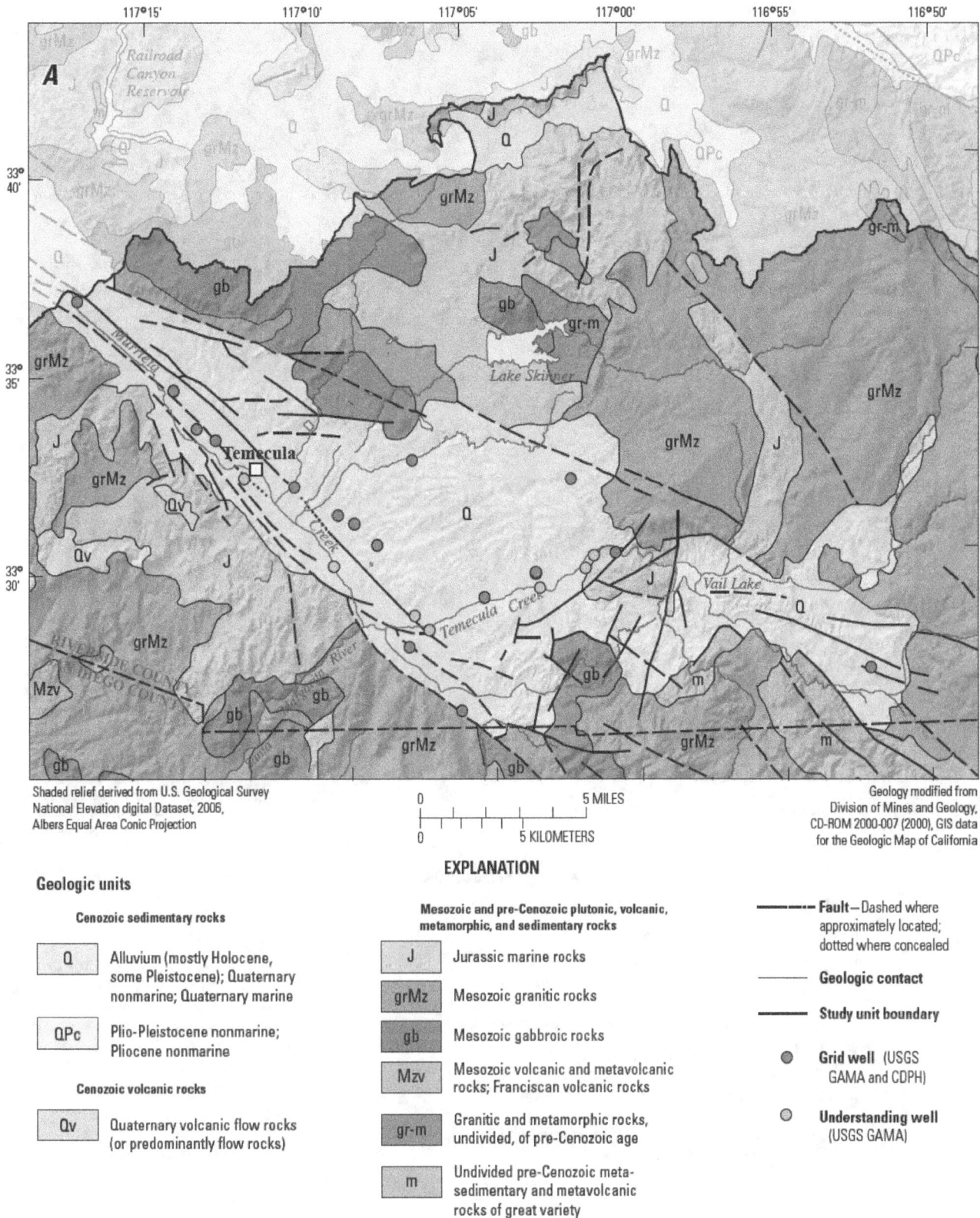

Shaded relief derived from U.S. Geological Survey
National Elevation digital Dataset, 2006,
Albers Equal Area Conic Projection

0   5 MILES

0   5 KILOMETERS

Geology modified from
Division of Mines and Geology,
CD-ROM 2000-007 (2000), GIS data
for the Geologic Map of California

EXPLANATION

Geologic units

Cenozoic sedimentary rocks

Q — Alluvium (mostly Holocene, some Pleistocene); Quaternary nonmarine; Quaternary marine

QPc — Plio-Pleistocene nonmarine; Pliocene nonmarine

Cenozoic volcanic rocks

Qv — Quaternary volcanic flow rocks (or predominantly flow rocks)

Mesozoic and pre-Cenozoic plutonic, volcanic, metamorphic, and sedimentary rocks

J — Jurassic marine rocks

grMz — Mesozoic granitic rocks

gb — Mesozoic gabbroic rocks

Mzv — Mesozoic volcanic and metavolcanic rocks; Franciscan volcanic rocks

gr-m — Granitic and metamorphic rocks, undivided, of pre-Cenozoic age

m — Undivided pre-Cenozoic metasedimentary and metavolcanic rocks of great variety

——··—— **Fault**—Dashed where approximately located; dotted where concealed

——— **Geologic contact**

——— **Study unit boundary**

● **Grid well** (USGS GAMA and CDPH)

○ **Understanding well** (USGS GAMA)

Figure 5A–C.   The geology of the San Diego Groundwater Ambient Monitoring and Assessment (GAMA) study unit and study areas: (A) Temecula Valley, (B) Warner Valley, (C) Alluvial Basins, and Hard Rock study areas.

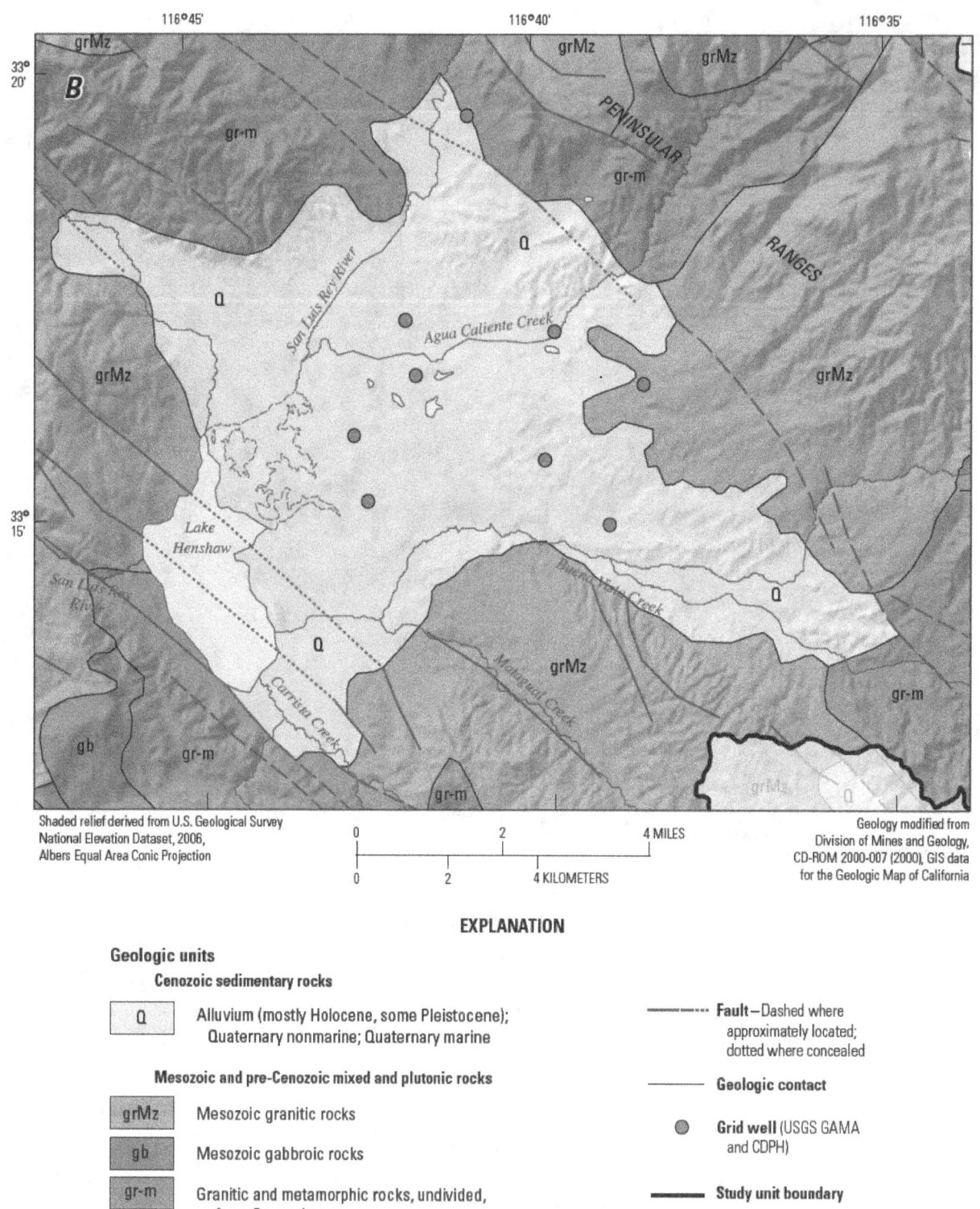

Shaded relief derived from U.S. Geological Survey
National Elevation Dataset, 2006,
Albers Equal Area Conic Projection

Geology modified from
Division of Mines and Geology,
CD-ROM 2000-007 (2000), GIS data
for the Geologic Map of California

0   2   4 MILES

0   2   4 KILOMETERS

EXPLANATION

Geologic units

Cenozoic sedimentary rocks

| Q | Alluvium (mostly Holocene, some Pleistocene); Quaternary nonmarine; Quaternary marine |

Mesozoic and pre-Cenozoic mixed and plutonic rocks

grMz	Mesozoic granitic rocks
gb	Mesozoic gabbroic rocks
gr-m	Granitic and metamorphic rocks, undivided, of pre-Cenozoic age

– – – – – –   **Fault**—Dashed where approximately located; dotted where concealed

————   **Geologic contact**

●   **Grid well** (USGS GAMA and CDPH)

▬▬▬▬   **Study unit boundary**

Figure 5*A–C.*—Continued

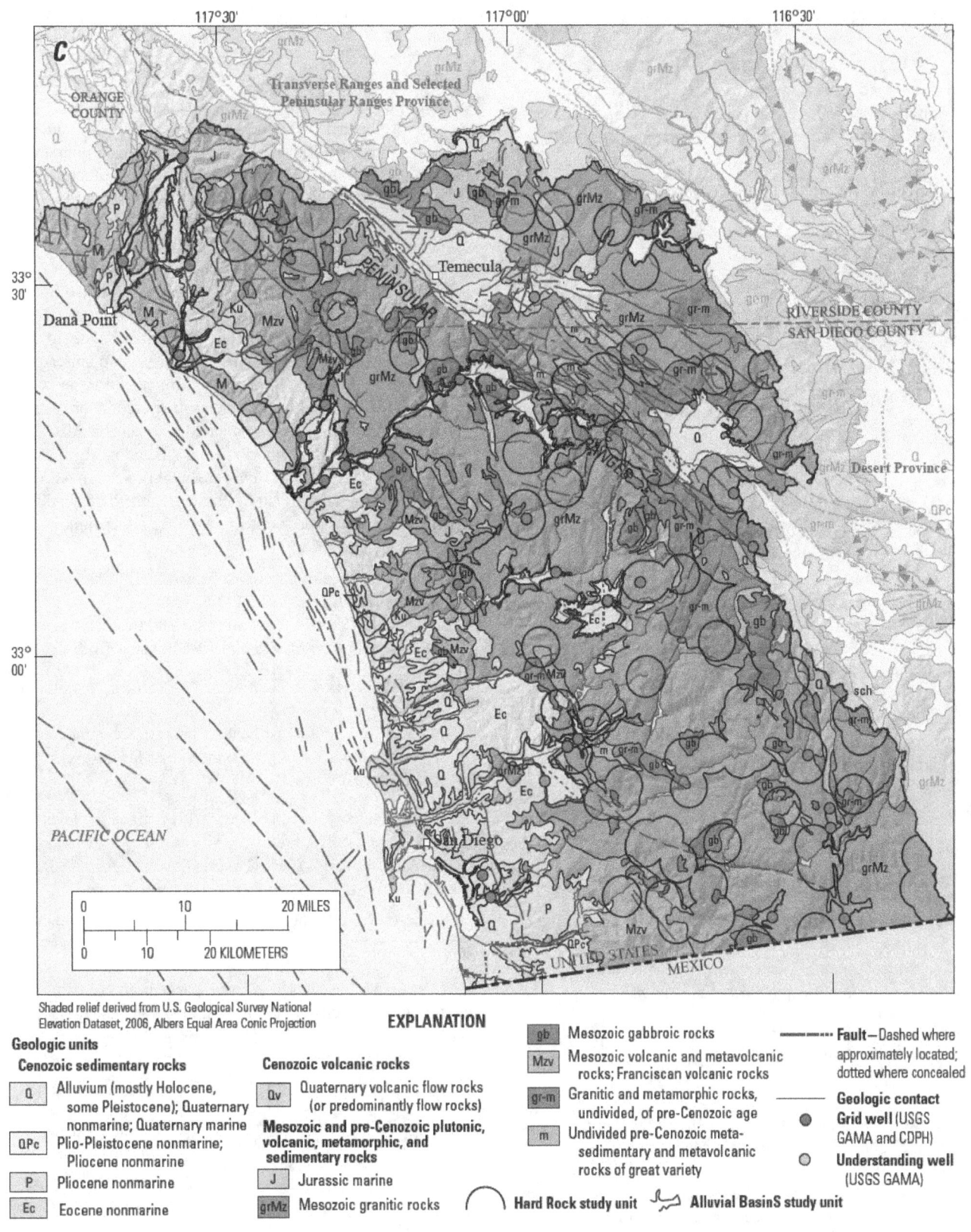

Shaded relief derived from U.S. Geological Survey National
Elevation Dataset, 2006, Albers Equal Area Conic Projection

EXPLANATION

Geologic units

Cenozoic sedimentary rocks

Q	Alluvium (mostly Holocene, some Pleistocene); Quaternary nonmarine; Quaternary marine
QPc	Plio-Pleistocene nonmarine; Pliocene nonmarine
P	Pliocene nonmarine
Ec	Eocene nonmarine

Cenozoic volcanic rocks

| Qv | Quaternary volcanic flow rocks (or predominantly flow rocks) |

Mesozoic and pre-Cenozoic plutonic, volcanic, metamorphic, and sedimentary rocks

| J | Jurassic marine |
| grMz | Mesozoic granitic rocks |

gb	Mesozoic gabbroic rocks
Mzv	Mesozoic volcanic and metavolcanic rocks; Franciscan volcanic rocks
gr-m	Granitic and metamorphic rocks, undivided, of pre-Cenozoic age
m	Undivided pre-Cenozoic meta-sedimentary and metavolcanic rocks of great variety

⌒ Hard Rock study unit   Alluvial BasinS study unit

------ **Fault**—Dashed where approximately located; dotted where concealed

—— **Geologic contact**

● **Grid well** (USGS GAMA and CDPH)

○ **Understanding well** (USGS GAMA)

Figure 5A–C.—Continued

Methods

The *status assessment* provides a spatially unbiased assessment of groundwater quality within in the primary aquifers, whereas the *understanding assessment* was designed to evaluate the natural and human factors that affect the groundwater quality of the San Diego study unit. The *status assessment* was conducted for each study area. This section describes the methods used for (1) defining groundwater quality, (2) assembling the datasets used for the *status assessment*, (3) determining which constituents warrant assessment, (4) calculating aquifer-scale proportions, and (5) analyzing statistics for the *understanding assessment*.

The primary metric for defining groundwater quality is *relative-concentration*, which references concentrations of constituents measured in groundwater to regulatory and non-regulatory benchmarks used to evaluate drinking water quality. Constituents are included or not included in the assessment on the basis of objective criteria by using these relative-concentrations. Groundwater-quality data collected by USGS-GAMA and data compiled in the CDPH database are used in the *status assessment*. Two statistical methods based on spatially unbiased equal-area grids are used to calculated aquifer-scale proportions of low, moderate, or high relative-concentrations: the "grid-based" method uses one value per cell to represent groundwater quality and the "spatially weighted" method uses many values per cell.

The CDPH database contains historical records from more than 25,000 wells, necessitating targeted retrievals to effectively access relevant water-quality data. The CDPH data were used in three ways in the *status assessment*: (1) to fill in gaps in the USGS data for the grid-based calculations of aquifer-scale proportions, (2) to identify constituents for inclusion in the assessment, and (3) to provide the majority of the data used in the spatially-weighted calculations of aquifer-scale proportions.

Relative-Concentrations and Water-Quality Benchmarks

Concentrations of constituents are presented as relative-concentrations in the *status assessment*:

$$Relative\text{-}concentration = \frac{Sample\ concentration}{Benchmark\ concentration}.$$

Relative-concentrations were used because they provide context for the measured concentrations in the sample: relative-concentrations less than 1 indicate sample concentrations less than the benchmark, and values greater than 1 indicate sample concentrations greater than the benchmark. The use of relative-concentrations also permits comparison on a single scale of constituents present at a wide range of concentration.

Toccalino and others (2004), Toccalino and Norman (2006), and Rowe and others (2007) previously used the ratio of measured sample concentration to the benchmark concentration (either maximum contaminant levels (MCLs) or Health-Based Screening Levels (HBSL)), and defined this ratio as the benchmark quotient. Relative-concentrations used in this report are equivalent to the benchmark quotient reported by Toccalino and others (2004) for constituents that have MCLs. However, HBSLs were not used in this report, as they are not currently used as benchmarks by California drinking-water regulatory agencies. Relative-concentrations can be computed only for constituents with water-quality benchmarks; therefore, constituents lacking water-quality benchmarks are not included in the *status assessment*.

Regulatory and non-regulatory benchmarks apply to treated water that is served to the consumer, not to untreated groundwater. However, to provide some context for the results, concentrations of constituents measured in the untreated groundwater were compared with benchmarks established by the U.S. Environmental Protection Agency (USEPA) and CDPH (U.S. Environmental Protection Agency, 2006; California Department of Public Health, 2008a, b). The benchmarks used for each constituent were selected in the following order of priority:

1. Regulatory, health-based CDPH and USEPA maximum contaminant levels (MCL-CA and MCL-US, respectively), USEPA action levels and treatment technique levels (AL-US and TT-US, respectively).

2. Non-regulatory CDPH and USEPA secondary maximum contaminant levels (SMCL-CA and SMCL-US, respectively). For constituents with both recommended and upper SMCL-CA levels, the values for the upper levels were used.

3. Non-regulatory, health based CDPH notification levels (NL-CA), USEPA lifetime health advisory levels (HAL-US), and USEPA risk-specific doses for 1:100,000 (RSD5-US).

Note that for constituents with multiple types of benchmarks, this hierarchy may not result in selection of the benchmark with the lowest concentration.

For ease of discussion, relative-concentrations of constituents were classified into low, moderate, and high categories:

Category	Relative-concentrations for organic constituents	Relative-concentrations for inorganic constituents
High	> 1	> 1
Moderate	> 0.1 and ≤ 1	> 0.5 and ≤ 1
Low	≤ 0.1	≤ 0.5

The boundary between "moderate" and "low" relative-concentrations was set at 0.1 for organic and special-interest constituents for consistency with other studies and reporting requirements (Toccalino and others, 2004; U.S. Environmental Protection Agency, 1998). For inorganic constituents, the boundary between "moderate" and "low" relative-concentrations was set at 0.5. A larger boundary value was used because in the San Diego study unit, and elsewhere in California (Landon and others, 2010), the naturally occurring inorganic constituents tend to be more prevalent in groundwater. Although more complex classifications could be devised based upon the properties and sources of individual constituents, use of a single moderate/low boundary value for each of the two major groups of constituents provided a consistent objective criteria for distinguishing constituents occurring at moderate rather than low concentrations.

Datasets for Status Assessment

USGS-Grid and -Understanding Wells

The primary data used for the grid-based calculations of aquifer-scale proportions of relative-concentrations were data from wells sampled by USGS-GAMA. Detailed descriptions of the methods used to identify wells for sampling are given in Wright and others (2005). Briefly, each study area was divided into equal-area grid cells that ranged in size and number from 10 4-mi^2 cells in the Warner Valley study area to 20 approximately 15-mi^2 cells in the Temecula Valley and Alluvial Basins study areas (fig. 6A–C). Because the Hard Rock study area was so large (850 mi^2), grids were configured to provide a sampling density of approximately one well per 85 mi^2 which equaled ten grid cells. The objective of the sampling design in the Hard Rock study area was to provide an initial reconnaissance of groundwater quality outside of CDWR-defined groundwater basins. Consequently the analyses from groundwater wells sampled in the Hard Rock study area were not included when calculating aquifer-scale proportions for constituents at the study unit level.

Within each grid cell, one well was randomly selected to represent the cell (Scott, 1990). It should be noted that some cells were divided into several sections because of geographic features (fig. 6A–C). Wells were selected from the population of wells in state-wide databases maintained by the CDPH and the USGS. USGS-grid wells in the San Diego study unit were numbered in the order of sample collection with the prefix varying by study area: the Temecula Valley study area (SDTEM), the Warner Valley study area (SDWARN), the Alluvial Basins study area (SDALLV), and the Hard Rock study area (SDHDRK). Grid well numbers in the San Diego study unit are not always sequential because some grid wells

have been re-designated as understanding wells subsequent to the publication of the San Diego study unit USGS Data Series report (table A1). Wells were redesignated in order to obtain a spatially distributed grid sampling-network that would meet the requirements of the *status assessment*.

The San Diego study unit contained a total of 60 grid cells, and the USGS sampled wells in 47 of those cells (USGS-grid wells) (fig. 6A–C). All 47 USGS-grid wells sampled in the San Diego study unit were PSWs that are listed in the CDPH water-quality database. Some grid cells could not be sampled because wells were not available, the wells were inoperable or the owner declined to participate in the program. However, if there was a well adjacent (≤ 1 km) to an empty grid cell, then the adjacent well was sampled and the water quality was used to represent the previously empty grid cell. Of the 20 grid cells in the Temecula Valley and Alluvial Basins study areas, 12 and 16 grid cells, respectively, were sampled or water-quality data was available from CDPH. In the Warner Valley, 9 of 10 grid cells were sampled and in the Hard Rock study area all 10 grid cells were sampled.

Eleven understanding wells were sampled for the purpose of understanding water quality changes along flow paths or in areas where historically little water-quality data were available. USGS-understanding (nonrandomized) wells were designated with the suffix FP for flow path wells and U for other understanding wells in addition to the regular GAMA ID. The understanding wells were not included in the grid-based characterization of water quality, but were used in the spatially weighted approach and were used to examine the effects of explanatory factors, such as land use, on water quality. An in-depth analysis of how water quality changes along flow paths in the San Diego study is not presented in this report.

Wells were sampled using a tiered analytical approach. All wells were sampled for a standard set of constituents, including VOCs, pesticides and pesticide degradates, stable isotopes of water, dissolved noble gases, and tritium (table 1). The standard set of constituents was termed the "fast" schedule. Wells on the "intermediate" schedule were sampled for all the constituents on the fast schedule, plus NDMA, perchlorate, potential waste-water indicators, and chromium species. Wells sampled on the "slow" schedule were sampled for all the constituents on the intermediate schedule, plus nutrients and dissolved organic carbon, major and minor ions, trace elements, arsenic and iron species, carbon isotopes, radon-222, radium isotopes, gross alpha and beta radiation, 1,4-dioxane and microbial constituents. Approximately 60 percent of the wells were sampled on a fast or intermediate schedule, and 40 percent were sampled on a slow schedule. Wells in areas of interest, such as along flow paths, or in places where water quality data were scarce, were given priority for slow schedule sampling.

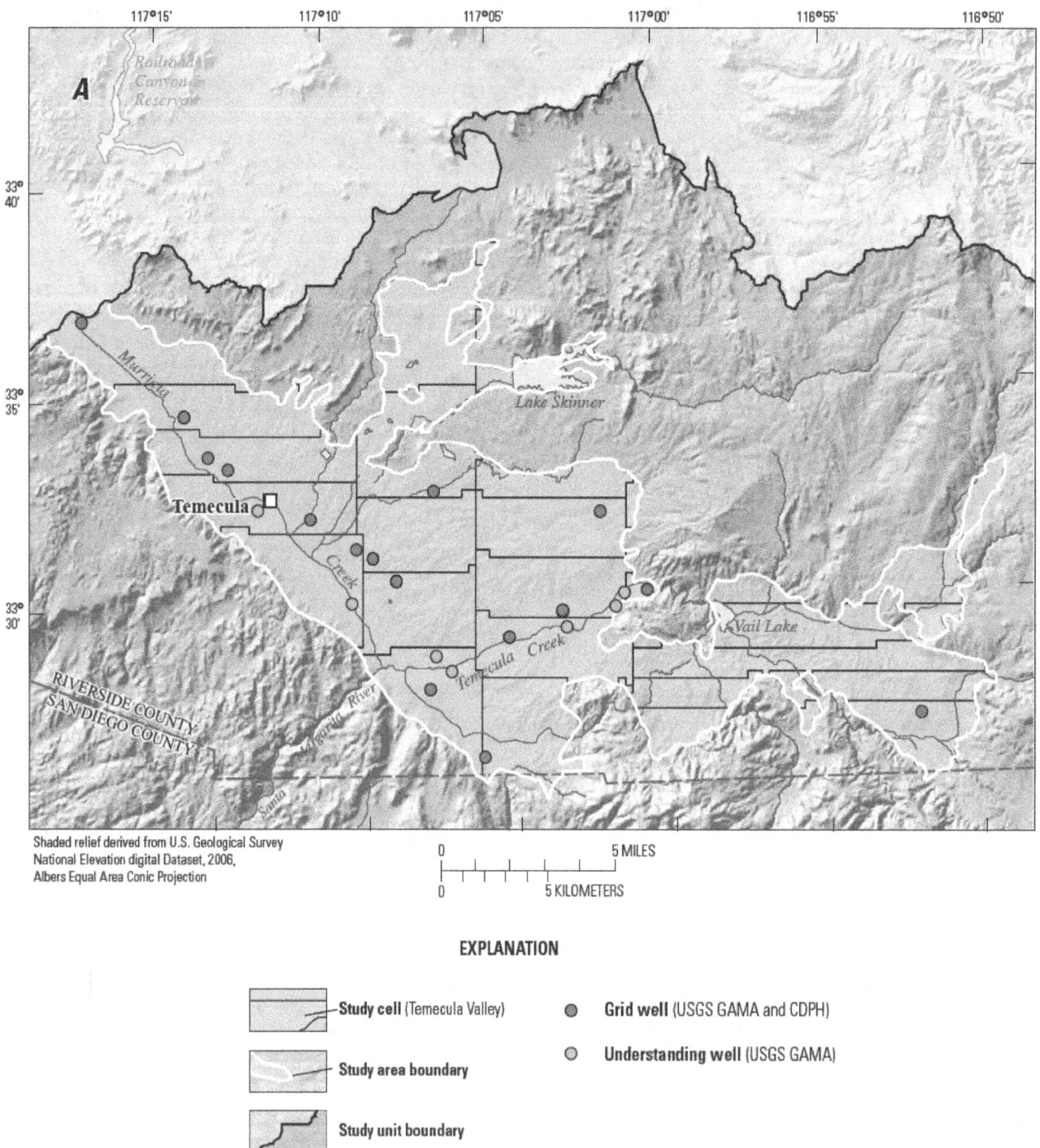

Shaded relief derived from U.S. Geological Survey
National Elevation digital Dataset, 2006,
Albers Equal Area Conic Projection

EXPLANATION

Study cell (Temecula Valley)   Grid well (USGS GAMA and CDPH)

Study area boundary   Understanding well (USGS GAMA)

Study unit boundary

Figure 6*A–C.*   Locations of grid cells, California Department of Public Health (CDPH) wells, and the USGS-grid and -understanding wells sampled during May to July, 2004 for the San Diego Groundwater Ambient Monitoring and Assessment (GAMA) study unit: (*A*) Temecula Valley, (*B*) Warner Valley, and (*C*) Alluvial Basins, and Hard Rock study areas.

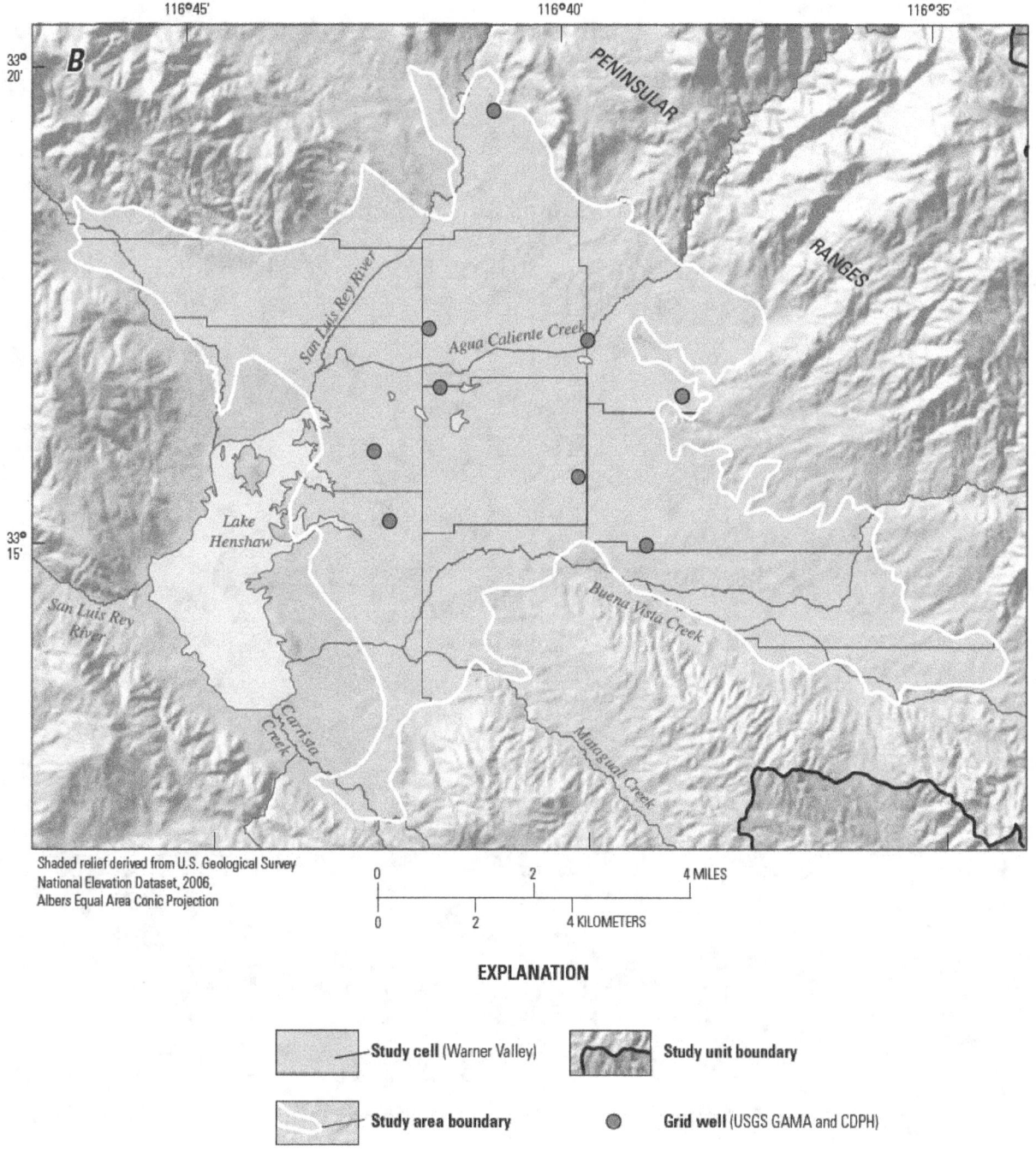

Shaded relief derived from U.S. Geological Survey
National Elevation Dataset, 2006,
Albers Equal Area Conic Projection

0   2   4 MILES

0   2   4 KILOMETERS

EXPLANATION

Study cell (Warner Valley)   Study unit boundary

Study area boundary   Grid well (USGS GAMA and CDPH)

Figure 6A–C.—Continued

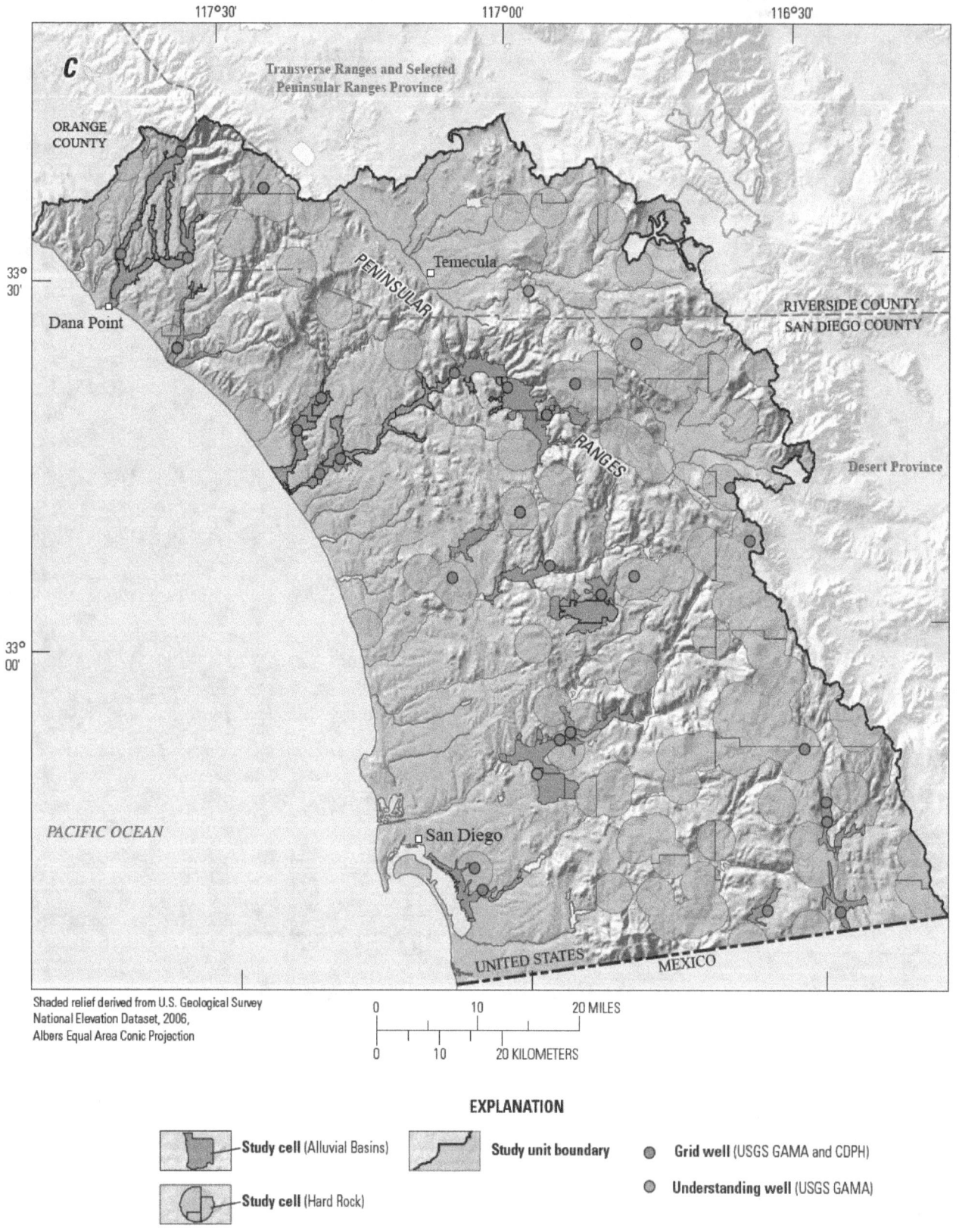

Shaded relief derived from U.S. Geological Survey
National Elevation Dataset, 2006,
Albers Equal Area Conic Projection

EXPLANATION

Study cell (Alluvial Basins)   Study unit boundary   ● Grid well (USGS GAMA and CDPH)

Study cell (Hard Rock)   ○ Understanding well (USGS GAMA)

Figure 6*A–C.*—Continued

Table 1. Constituent class and numbers of constituents and wells sampled for each analytical group in the San Diego Groundwater Ambient Monitoring and Assessment (GAMA) study unit, California, May 17–July 29, 2004.

[NDMA, *N*-Nitrosodimethylamine; USGS, U.S. Geological Survey]

| | Sampling schedule | | |
| | Fast | Intermediate | Slow |
Well summary	**Number of wells**		
Total number of wells	8	26	24
Number of grid wells sampled	6	22	19
Number of understanding wells sampled	2	4	5
Analyte Groups[1]	**Number of constituents**		
Specific conductance and temperature	2	2	2
Volatile organic compounds (VOCs)	88	88	88
Pesticides and degradates	64	64	64
Noble gases and tritium[2]	7	7	7
Stable isotopes of hydrogen and oxygen	2	2	2
Potential waste-water indicators[3]		48	48
Pharmaceuticals[4]		16	16
Perchlorate and NDMA		2	2
Chromium species		2	2
Tritium[5]		1	1
pH, dissolved oxygen, alkalinity, turbidity			4
Polar pesticides and degradates[6]			59
1,4-Dioxane			1
Nutrients and dissolved organic carbon			6
Major and minor ions, and trace elements			36
Arsenic and iron species			4
Carbon isotopes			2
Radon-222			1
Radium isotopes			2
Gross alpha and beta radioactivity			4
Microbial constituents			4
Sum of constituents for each schedule:	163	232	355

[1]Not all analyte groups or analytes are discussed in the report.

[2]Analyzed at Lawrence Livermore National Laboratory, Livermore, California.

[3]Counts do not include analytes in common with VOCs, pesticides and degradates, pharmaceuticals or polar pesticides and degradates. Wastewater data is not used for assessment of status or understanding in this report.

[4]Pharmaceutical data is not used for assessment of status or understanding in this report.

[5]Analyzed at USGS Stable Isotope and Tritium Laboratory, Menlo Park, California.

[6]Counts do not include analytes in common with pesticides and degradates.

CDPH Grid Wells

The four study areas were divided into 60 grid cells, out of which no USGS-grid wells were available for 13 cells; USGS-grid wells were available for 28 cells but no USGS data for major ions, trace elements, nutrients, and radiochemical constituents were available. Data from the CDPH database were used to provide missing inorganic and radiochemical data. CDPH wells with data for the most recent 3 years available at the time of sampling (July 30, 2001 through July 29, 2004) were considered. If more than one analysis for a constituent was available in the 3-year interval for a well, then the most recent data were selected.

The decision tree used to identify suitable data from CDPH wells is described in appendix A. Briefly, the first choice was to use CDPH data from the same well sampled by the USGS (USGS-grid well). In this case, "DG" was added to the well's GAMA ID to signify that it was a well sampled by the USGS but also whose data were supplemented from the CDPH database (fig. A1A–C; table A1). If all the needed data for the DG well were not available, then a second well in the cell was randomly selected from the subset of CDPH wells with data and a new identification with "DPH" and a new number was assigned to that well. The combination of the USGS-grid wells and the CDPH-grid wells produced a grid-well network covering 54 of the 60 grid cells in the San Diego study unit.

Note that the CDPH database generally did not contain data for all of the missing inorganic constituents at every CDPH-grid well; therefore, the number of wells used for the grid-based assessment was different for different inorganic constituents (table 2). Although other organizations also collect water-quality data, the CDPH data is the only Statewide database of groundwater-chemistry data available for comprehensive analysis.

CDPH data were not used to supplement USGS-grid well data for VOCs, pesticides, or perchlorate for the grid-based status assessment. A larger number of VOCs and pesticide compounds are analyzed for the USGS-GAMA Program than are available from CDPH. USGS-GAMA collected data for 88 VOCs plus 64 pesticides and pesticide degradates at every well in the San Diego study unit (table 1). In addition, method detection limits for USGS-GAMA analyses of organic constituents typically were one to two orders of magnitude lower than the reporting limits for analyses compiled by CDPH (table 3).

Table 2. Inorganic constituents and number of grid wells per constituent, San Diego Groundwater Ambient Monitoring and Assessment (GAMA) study unit, May–July 2004.

[CDPH, California Department of Public Health; N, nitrogen; SMCL, Secondary Maximum Contaminant Level; HBB, Health Based Benchmark (including all benchmark types except SMCL); USGS, U.S. Geological Survey]

Constituent type	Constituent	Number of grid wells	
		Sampled by USGS GAMA	Selected from CDPH
Major element—SMCL			
	Chloride	19	14
	Sulfate	19	14
	Total dissolved solids	19	16
Minor element—HBB			
	Fluoride	19	14
Nutrient—HBB			
	Nitrite-N	19	15
	Ammonia-N	19	0
	Nitrate-N	19	18
Radioactive—HBB			
	Gross alpha radioactivity	19	16
	Gross beta radioactivity	19	4
	Ra226+228	19	0
	Rn222	19	0
	Uranium	19	8
Trace element—HBB			
	Aluminum	19	15
	Antimony	19	14
	Arsenic	19	15
	Barium	19	15
	Beryllium	19	14
	Boron	19	15
	Cadmium	19	15
	Chromium	19	14
	Copper	19	16
	Lead	19	14
	Mercury	19	15
	Nickel	19	14
	Selenium	19	15
	Strontium	19	0
	Thallium	19	14
	Vanadium	19	14
Trace element—SMCL			
	Iron	19	15
	Manganese	19	15
	Silver	19	15
	Zinc	19	15

Table 3. Comparison of number of compounds and median method detection limit or laboratory reporting levels by type of constituent for data stored in the California Department of Public Health database and data collected for the San Diego Groundwater Ambient Monitoring and Assessment (GAMA) study unit, California, May 17–July 29, 2004.

[CDPH, California Department of Public Health; MDL, method detection limit; LRL, laboratory reporting level; MRL, method reporting level; mg/L, milligrams per liter; µg/L, micrograms per liter; nc, not collected; pCi/L, picocuries per liter]

Constituent type	CDPH		GAMA	
	Number ofcompounds	MDL	Number ofcompounds	MedianLRL/MRL
Volatile organic compounds (µg/L)	61	0.5	88	0.06
Pesticides and degradates (µg/L)	27	2	123	0.019
Nutrients, major and minor ions (mg/L)	4	0.4	17	0.06
Trace elements (µg/L)	20	8	25	0.12
Radioactive constituents (SSMDC)[1] (pCi/L)	5	1	8	0.54
Perchlorate (µg/L)	1	4	1	0.5
1,4-Dioxane (µg/L)	1	3	1	2
N-Nitrosodimethylamine (NDMA) (µg/L)	nc	nc	1	0.002
Pharmaceutical constituents (µg/L)	nc	nc	16	0.021

[1] Value reported for the median LRL/MRL is a median sample-specific critical level for eight radioactive constituents collected and analyzed by GAMA.

Additional Data Used For Spatially Weighted Calculation

The spatially weighted calculations of aquifer-scale proportions of relative-concentrations used data from the USGS-grid wells, additional wells sampled by USGS-GAMA (understanding wells), and all wells in the CDPH database with water-quality data during the 3-year interval July 30, 2001, through July 29, 2004. For wells with both USGS and CDPH data, only the USGS data were used.

Identification of Constituents for Status Assessment

Three criteria were used to identify constituents for additional evaluation in the status assessment of groundwater in the San Diego study:

1. Constituents present at high or moderate relative-concentrations in the CDPH database within the 3-year interval;

2. Constituents present at high or moderate relative-concentrations in the USGS-grid wells or USGS-understanding well;

3. Organic constituents with study unit detection frequencies greater than 10 percent in the USGS-grid well dataset.

These criteria identified 11 organic and special-interest constituents and 26 inorganic constituents for additional evaluation in the *status assessment* (table 4). An additional 23 organic constituents and 20 inorganic constituents were detected by USGS-GAMA, and are not included for further analysis in the *status assessment* because they either have no established benchmarks (table 5), or were only detected at low relative-concentrations.

The CDPH database also was used to identify constituents that have been reported at high relative-concentrations historically, but not at the time of this study. The historical period was defined as from the earliest record maintained in the CDPH electronic database within the period May 1983 to June, 2001. Constituents may be historically high, but not currently high, because of improvement in groundwater quality with time or abandonment of wells with high relative-concentrations. Historically high constituents that do not otherwise meet the criteria for inclusion in the *status assessment* are not considered representative of potential groundwater-quality concerns in the study unit from 2001 to 2004. For the San Diego study unit, there were six historically high constituents (table 6).

Table 4. Aquifer proportions from grid-based and spatially weighted methods for constituents detected in the Alluvial Fill study areas (Temecula Valley, Warner Valley, and Alluvial Basins) (1) with high relative-concentrations reported in the California Department of Public Health (CDPH) database during July 30, 2001–July 29 2004, or (2) with moderate or high relative-concentrations in samples collected from grid wells during May–July 2004, or (3) with organic or special-interest constituents detected in more than 10 percent of samples collected from grid wells during May–July 2004, San Diego Groundwater Ambient Monitoring and Assessment (GAMA) study unit, California.

[Grid-based aquifer-scale proportions for organic constituents are based on samples collected by the U.S. Geological Survey from 47 grid wells during May–July 2004. Spatially weighted aquifer-scale proportions are based on CDPH data for July 30, 2001–July 29, 2004, combined with grid-well and understanding-well data. High, concentrations greater than benchmark; moderate, concentrations less than benchmark and greater than or equal to 0.1 of benchmark for inorganic constituents; low, concentrations less than 0.1 of benchmark for organic constituents or 0.5 of benchmark for inorganic constituents. HAL-US, USEPA lifetime health advisory level; MCL-US, USEPA maximum contaminant level; MCL-CA, CDPH maximum contaminant level; NL-CA, CDPH notification level; SMCL-CA, CDPH secondary maximum contaminant level; USEPA, U.S. Environmental Protection Agency; CDPH, California Department of Public Health; µg/L, microgram per liter; pCi/L, picocurie per liter; ns, not sampled; mg/L, milligram per liter]

Constituent	Threshold type	Threshold value	Threshold units	Grid-based aquifer-scale proportion[1]				Raw detection frequency[2]			Spatially weighted aquifer-scale proportion[1,2]	
				Number of wells	High (percent)	Moderate (percent)	Low (percent)	Number of wells	Number of wells with high values	Raw detection frequency (percent)	Number of cells	High (percent)
Trace elements												
Vanadium	NL-CA	50	µg/L	28	7.5	28.5	64.0	117	12	10.3	32	9.8
Arsenic	MCL-US	10	µg/L	29	4.1	2.8	93.1	117	4	3.4	32	3.8
Boron	NL-CA	1,000	µg/L	28	4.1	0.0	95.9	117	2	1.7	32	1.3
Antimony	MCL-US	6	µg/L	28	3.4	0.0	96.6	106	1	0.9	31	3.2
Selenium	MCL-US	50	µg/L	29	0.0	3.2	96.8	117	0	0.0	32	0.0
Fluoride	MCL-CA	2	mg/L	28	[3]0.7	2.8	97.2	115	1	0.9	31	0.7
Aluminum	MCL-CA	1,000	µg/L	29	0.0	0.0	100.0	117	0	0.0	32	0.0
Thallium	MCL-US	2	µg/L	28	0.0	0.0	100.0	106	0	0.0	32	0.0
Chromium	MCL-CA	50	µg/L	28	0.0	0.0	100.0	109	0	0.0	32	0.0
Lead	AL-US	15	µg/L	28	0.0	0.0	100.0	114	0	0.0	31	0.0
Nickel	MCL-CA	100	µg/L	28	0.0	0.0	100.0	105	0	0.0	32	0.0
Cadmium	MCL-US	5	µg/L	29	0.0	0.0	100.0	117	0	0.0	32	0.0
Mercury	MCL-US	2	µg/L	29	0.0	0.0	100.0	117	0	0.0	32	0.0
Radioactive constituents												
Gross-alpha radioactivity	MCL-US	15	pCi/L	29	3.2	10.5	86.3	115	4	3.5	30	3.4
Uranium	MCL-US	30	µg/L	22	0.0	13.1	86.9	53	1	1.9	26	0.9
Radon-222	Proposed AMCL-US	4,000	pCi/L	15	0.0	0.0	100.0	21	0	0.0	18	0.0
Radium-228	MCL-US	5	pCi/L	14	0.0	0.0	100.0	34	0	0.0	20	0.0
Gross-beta radioactivity	MCL-CA	50	pCi/L	19	0.0	0.0	100.0	47	0	0.0	22	0.0
Nutrients												
Nitrate as nitrogen	MCL-US	10	mg/L	31	3.4	6.8	89.8	148	7	4.7	36	3.4
Nitrite, as nitrogen	MCL-US	1	mg/L	28	0.0	0.0	100.0	127	0	0.0	33	0.0

Table 4. Aquifer proportions from grid-based and spatially weighted methods for constituents detected in the Alluvial Fill study areas (Temecula Valley, Warner Valley, and Alluvial Basins) (1) with high relative-concentrations reported in the California Department of Public Health (CDPH) database during July 30, 2001–July 29 2004, or (2) with moderate or high relative-concentrations in samples collected from grid wells during May–July 2004, or (3) with organic or special-interest constituents detected in more than 10 percent of samples collected from grid wells during May–July 2004, San Diego Groundwater Ambient Monitoring and Assessment (GAMA) study unit, California.—Continued.

[Grid-based aquifer-scale proportions for organic constituents are based on samples collected by the U.S. Geological Survey from 47 grid wells during May–July 2004. Spatially weighted aquifer-scale proportions are based on CDPH data for July 30, 2001–July 29, 2004, combined with grid-well and understanding-well data. High, concentrations greater than benchmark; moderate, concentrations less than benchmark and greater than or equal to 0.1 of benchmark for organic constituents or 0.5 of benchmark for inorganic constituents; low, concentrations less than 0.1 of benchmark for organic constituents or 0.5 of benchmark for inorganic constituents. HAL-US, USEPA lifetime health advisory level; MCL-US, USEPA maximum contaminant level; MCL-CA, CDPH maximum contaminant level; NL-CA, CDPH notification level; SMCL-CA, CDPH secondary maximum contaminant level; USEPA, U.S. Environmental Protection Agency; CDPH, California Department of Public Health; μg/L, microgram per liter; ns, not sampled; mg/L, milligram per liter; pCi/L, picocurie per liter]

Constituent	Threshold type	Threshold value	Threshold units	Grid-based aquifer-scale proportion[1]				Raw detection frequency[2]			Spatially weighted aquifer-scale proportion[1,2]	
				Number of wells	High (percent)	Moderate (percent)	Low (percent)	Number of wells	Number of wells with high values	Raw detection frequency (percent)	Number of cells	High (percent)
Major and minor elements (SMCLs)												
Total dissolved solids (TDS)	SMCL-CA	1,000	mg/L	29	13.7	31.2	55.1	119	15	12.6	34	13.0
Manganese	SMCL-CA	50	μg/L	28	13.7	3.4	82.9	120	32	26.7	32	21.2
Iron	SMCL-CA	300	μg/L	28	6.9	0.0	93.1	120	10	8.3	32	6.5
Chloride	SMCL-CA	500	mg/L	28	3.4	10.3	86.3	117	5	4.3	32	2.7
Sulfate	SMCL-CA	500	mg/L	28	3.4	6.9	89.7	117	2	1.7	32	3.7
Zinc	SMCL-CA	5,000	μg/L	29	0.0	0.0	100.0	114	0	0.0	32	0.0
Gasoline components												
MTBE	MCL-CA	13	μg/L	37	3.0	0.0	97.0	113	2	1.8	39	1.4
Benzene	MCL-CA	1	μg/L	37	0.0	0.0	100.0	113	0	0.0	39	0.0
Trihalomethanes												
Chloroform	MCL-US	[4]80	μg/L	37	0.0	0.0	100.0	112	0	0.0	38	0.0
Solvents												
Tetrachloroethylene	MCL-US	5	μg/L	37	0.0	0.0	100.0	113	0	0.0	39	0.0
Carbon tetrachloride	MCL-CA	0.5	μg/L	37	0.0	0.0	100.0	113	0	0.0	39	0.0
Trichloroethylene	MCL-US	5	μg/L	37	0.0	0.0	100.0	113	0	0.0	39	0.0
1,2-Dichloropropane	MCL-US	5	μg/L	37	0.0	3.0	97.0	113	0	0.0	39	0.0
Herbicides												
Prometon	HAL-US	100	μg/L	37	0.0	0.0	100.0	38	0	0.0	37	0.0
Simazine	MCL-US	4	μg/L	37	0.0	0.0	100.0	101	0	0.0	37	0.0
Atrazine	MCL-CA	1	μg/L	37	0.0	0.0	100.0	101	0	0.0	37	0.0

Table 4. Aquifer proportions from grid-based and spatially weighted methods for constituents detected in the Alluvial Fill study areas (Temecula Valley, Warner Valley, and Alluvial Basins) (1) with high relative-concentrations reported in the California Department of Public Health (CDPH) database during July 30, 2001–July 29 2004, or (2) with moderate or high relative-concentrations in samples collected from grid wells during May–July 2004, or (3) with organic or special-interest constituents detected in more than 10 percent of samples collected from grid wells during May–July 2004, San Diego Groundwater Ambient Monitoring and Assessment (GAMA) study unit, California.—Continued.

[Grid-based aquifer-scale proportions are based on samples collected by the U.S. Geological Survey from 47 grid wells during May–July 2004. Spatially weighted aquifer-scale proportions for organic constituents are based on samples collected by the U.S. Geological Survey from 47 grid wells during May–July 2004. Spatially weighted aquifer-scale proportions are based on CDPH data for July 30, 2001–July 29, 2004, combined with grid-well and understanding-well data. High, concentrations greater than benchmark; moderate, concentrations less than benchmark and greater than or equal to 0.1 of benchmark for organic constituents or 0.5 of benchmark for inorganic constituents; low, concentrations less than 0.1 of benchmark for organic constituents or 0.5 of benchmark for inorganic constituents. HAL-US, USEPA lifetime health advisory level; MCL-US, USEPA maximum contaminant level; MCL-CA, CDPH maximum contaminant level; NL-CA, CDPH notification level; SMCL-CA, CDPH secondary maximum contaminant level; USEPA, U.S. Environmental Protection Agency; CDPH, California Department of Public Health; µg/L, microgram per liter; ns, not sampled; mg/L, milligram per liter; pCi/L, picocurie per liter]

Constituent	Threshold type	Threshold value	Threshold units	Grid-based aquifer-scale proportion[1]				Raw detection frequency[2]			Spatially weighted aquifer-scale proportion[1,2]	
				Number of wells	High (percent)	Moderate (percent)	Low (percent)	Number of wells	Number of wells with high values	Raw detection frequency (percent)	Number of cells	High (percent)
Constituent of special interest												
Perchlorate	MCL-CA	6	µg/L	32	[3]0.2	36.3	63.7	103	1	0.9	38	0.2

[1] Alluvial Fill study area aquifer-scale proportion is calculated by summing the area weighted average for each individual study area except the Hard Rock. Area-weighted values for each study area are: Temecula Valley = 0.41, Warner Valley = 0.11, Alluvial Basins = 0.48. Aquifer-scale proportions will not sum to 100 if a spatially-weighted value is used.

[2] Based on most recent analysis for each CDPH well during July 30, 2001–July 29, 2004, combined with GAMA grid-based data.

[3] Spatially-weighted value. Aquifer-scale proportions will not sum to 100 if a spatially-weighted value is used.

The MCL-US threshold for trihalomethanes is the sum of chloroform, bromoform, bromodichloromethane, and dibromochloromethane.

Table 5. Number of constituents analyzed and detected by benchmark and constituent type, San Diego Groundwater Ambient Monitoring and Assessment (GAMA) study unit grid wells, California, May 17 to July 29, 2004.

[VOCs, volatile organic compounds; NWQL, National Water Quality Laboratory; USEPA, U.S. Environmental Protection Agency; CDPH, California Department of Public Health; MCL, USEPA or CDPH Maximum Contaminant Level; HAL, USEPA Health Advisory Level; NL, CDPH Notification Level; RSD5, USEPA Risk Specific Dose at 10^{-5}; AL, USEPA Action Level; SMCL, USEPA or CDPH Secondary Maximum Contaminant Level]

Groups of organic constituents

Benchmark type	Sum organic and special-interest constituents		VOCs + gasoline oxygenates		Pesticides and degradates (NWQL Schedule 2003)[1]		Polar pesticides and degradates (NWQL Schedule 2060)[1]		Special-interest constituents	
	Analyzed	Detected	Analyzed	Detected	Analyzed	Detected	Analyzed	Detected	Analyzed	Detected
MCL	46	16	32	12	3	2	10	1	1	1
HAL	31	6	7	0	14	5	10	1	0	0
NL	16	0	14	0	0	0	0	0	2	0
RSD5	8	1	4	0	3	0	1	0	0	0
AL	0	0	0	0	0	0	0	0	0	0
SMCL	0	0	0	0	0	0	0	0	0	0
None	117	11	31	0	44	10	42	1	0	0
Total	218	34	88	12	64	17	63	4	3	1

Groups of inorganic constituents

Benchmark type	Sum of inorganic constituents		Major and minor ion		Nutrients		Trace elements		Radioactive constituents	
	Analyzed	Detected	Analyzed	Detected	Analyzed	Detected	Analyzed	Detected	Analyzed	Detected
MCL	23	21	1	1	2	2	12	10	8	8
HAL	4	4	0	0	1	1	3	3	0	0
NL	2	2	0	0	0	0	2	2	0	0
RSD5	0	0	0	0	0	0	0	0	0	0
AL	2	2	0	0	0	0	2	2	0	0
SMCL	6	6	3	3	0	0	3	3	0	0
None	13	13	6	6	3	3	3	3	1	1
Total	50	48	10	10	6	6	25	23	9	9

Organic, special interest, inorganic total	
268	82

[1]There are four overlapping compounds between NWQL Schedules 2003 and 2060.

Table 6. Constituents with one or more concentrations above health-based benchmarks for the period of May 1983 to June 2001, based on the California Department of Public Health data for public-supply wells in the San Diego Groundwater Ambient Monitoring and Assessment (GAMA) study unit, California.

Constituent	Number of wells with analyses	Total number of analyses	Total number of analyses above threshold	Number of wells with at least one high analysis	Date of most recent concentration above a health-based benchmark
Trace elements					
Chromium	230	843	1	1	11-29-1989
Cadmium	246	843	3	3	05-22-1990
Mercury	248	859	1	1	02-26-1992
Radioactive constituents					
Gross-beta radioactivity	77	80	1	1	07-05-1995
Radium 226	41	109	4	3	06-26-1996
Solvents					
Tetrachloroethylene	269	1,173	19	2	10-10-2000
Trichloroethylene	270	1,173	25	2	07-10-2000
1,2-Dichloropropane	243	1,108	2	1	03-27-1995

Calculation of Aquifer-Scale Proportions

The *status assessment* is intended to characterize the quality of groundwater resources within the primary aquifers of the San Diego study unit. The primary aquifers are defined by the depth intervals over which wells listed in the CDPH database are perforated. The use of the term "primary aquifers" does not imply a discrete aquifer unit. In most groundwater basins, municipal and community supply wells generally are perforated at greater depths than are domestic wells. Thus, because domestic wells are not listed in the CDPH database, the primary aquifers generally correspond to the portion of the aquifer system tapped by municipal and community supply wells. All wells used in the *status assessment* in the San Diego study unit are listed in the CDPH database, and are therefore classified as municipal and community drinking-water supply wells.

Two statistical approaches, grid-based and spatially weighted (Belitz and others, 2010), were applied to evaluate the proportions of the primary aquifers in the San Diego study unit with high, moderate, and low relative-concentrations of constituents. For ease of discussion, these proportions are referred to as "high, moderate, and low aquifer-scale proportions." Calculations of aquifer-scale proportions were made for individual constituents meeting the criteria for additional evaluation in the *status assessment* and for classes of constituents. Classes of constituents with health-based benchmarks included: trace elements, radioactive constituents, nutrients, VOCs, and pesticides. Aquifer-scale proportions were also calculated for the following constituents having aesthetic (SMCL) benchmarks: manganese, total dissolved solids, iron, chloride, sulfate, and zinc.

The grid-based calculation uses the grid-well dataset assembled from the USGS- and CDPH-grid wells. For each constituent the high aquifer-scale proportion for a study area was calculated by dividing the number of cells (wells) represented by a high value for that constituent by the total number of grid cells with data for that constituent. The high aquifer-scale proportions at the study-unit scale were then calculated by first multiplying the study-area aquifer-scale proportion by an area-weighted correction factor, and then summing the high aquifer-scale proportions for all the study areas. An area-weighted correction factor was needed because the study areas are not the same size (fig. 6A–C). Moderate and low aquifer-scale proportions were calculated using the same approach as the calculations for the high aquifer-scale proportions. A more detailed discussion of the calculation used for aquifer-scale proportion is located in appendix B.

The grid-based estimate is spatially unbiased; however, this approach may not detect constituents that are present at high relative-concentrations in small proportions of the primary aquifers. The spatially weighted calculation uses all CDPH wells in the study unit (most recent analysis during the current period from July 30, 2001–July 29, 2004), USGS-grid wells, and USGS-understanding wells to represent the primary aquifers. By using the spatially weighted approach, the proportion of high relative-concentrations for the primary aquifers for each constituent was computed by (1) computing the proportion of wells with high relative-concentrations in each grid cell and (2) averaging together the grid-cell proportions computed in step (1) (Isaaks and Srivastava, 1989; Belitz and others, 2010). Similar procedures were used to calculate the proportions of the aquifer with moderate and low relative-concentrations of constituents. The resulting proportions are spatially unbiased (Isaaks and Srivastava, 1989; Belitz and others, 2010). Confidence intervals for spatially weighted detection frequencies of high relative-concentrations are not described in this report.

In addition, for each constituent, the raw detection frequencies of high and moderate values for individual constituents were calculated by using the same dataset as used for the spatially weighted calculations. However, raw detection frequencies are not spatially unbiased because the wells in the CDPH database are not uniformly distributed. For example, if a constituent were present at high relative-concentrations in a small region of the aquifer with a high density of wells, then the raw detection frequency of high values would be greater than the high aquifer-scale proportion. Raw detection frequencies are provided for reference but were not used to assess aquifer-scale proportions (see appendix B for details of statistical methods).

The grid-based high aquifer-scale proportions were used to represent proportions in the primary aquifers unless the grid-base high aquifer-scale proportion was zero and the spatially weighted proportion was non-zero, and then the spatially weighted result was used. This situation can arise when the relative-concentration of a constituent is high in a small fraction of the primary aquifers. The grid-based moderate and low proportions were used in most cases because the reporting limits for many organic constituents and some inorganic constituents in the CDPH database were higher than the boundary between the moderate and low categories. However, if the grid-based moderate proportion was zero and the spatially weighted proportion non-zero, then the spatially weighed value was used..

Understanding-Assessment Methods

Explanatory factors, including land use, well depth, depth to the top of the uppermost open interval, classified groundwater age, and redox conditions (see appendix C for more details), were analyzed in relation to constituents of interest for the *understanding assessment* in order to establish context for physical and chemical processes. Statistical tests were used to identify significant correlations between the constituents of interest and potential explanatory factors. Significant correlations for explanatory factors influencing water quality are shown in the figures.

The wells included in the *understanding assessment* include USGS-grid and CDPH-grid well and USGS-understanding wells. CDPH-other wells were not used in the *understanding assessment* because age tracer, dissolved oxygen, and sometimes well construction data were not available. For different potential explanatory variables, correlations were tested by using either the set of grid plus understanding wells or grid wells only. Because the USGS-understanding wells were not randomly selected on a spatially distributed grid, these wells were excluded from analyses of relations of water quality to areally-distributed variables (land use) to avoid areal-clustering bias. However, USGS-understanding wells were included in analyses of relations between constituents and the vertically distributed explanatory variables depth, classified groundwater age, and oxidation-reduction characteristics in order to have data spanning a sufficient range of variables to identify relations.

For inorganic constituents to be discussed in the *understanding assessment*, they must have been detected at high relative-concentrations in greater than or equal to 2 percent of the aquifer (based on non area-weighted detections for all study areas) For organic and special-interest constituents to be discussed in the understanding assessment, a constituent needs to be detected at a high or moderate relative-concentration, or detected in greater than or equal to 10 percent of grid wells (based on detections that were not area-weighted) regardless of concentration

Statistical Analysis

Nonparametric statistical methods were used to test the significance of correlations between water-quality variables and potential explanatory factors. Nonparametric statistics are robust techniques that generally are not affected by outliers and do not require that the data follow any particular distribution (Helsel and Hirsch, 2002). The significance level (p) used to test hypotheses for this report was compared to a threshold value (α) of 5 percent ($\alpha = 0.05$) to evaluate whether the relation was statistically significant ($p < \alpha$). Correlations were investigated using Spearman's method to calculate the rank-order correlation coefficient (ρ) between continuous variables. The values of ρ can range from +1.0 (perfect positive correlation) to 0.0 (no correlation) to -1.0 (perfect negative correlation).

The Wilcoxon rank-sum test was used to evaluate the correlation between water quality and categorical explanatory factors: for example, groundwater age (modern, mixed, or pre-modern), redox conditions (oxic, mixed, or anoxic/suboxic), and land-use classification (natural, agricultural, urban, or mixed). The Wilcoxon rank-sum test can be used to compare two independent populations (data groups or categories) to determine whether one population contains larger values than the other (Helsel and Hirsch, 2002). The null hypothesis for the Wilcoxon rank sum test is that there is no significant difference between the values of the two independent data groups being tested. The Wilcoxon rank sum test was used for multiple comparisons of two independent groups rather than the multiple-stage Kruskal-Wallis test for identifying differences between three or more groups, although a set of Wilcoxon rank sum tests is more likely to falsely indicate a significant difference between groups than the Kruskal-Wallis test (Helsel and Hirsch, 2002). However, given the potentially large and variable number of differences to be evaluated, the Wilcoxon rank sum test was selected as a consistent and practical direct test of differences. Because of the small sample size, the exact distribution with continuity correction also was applied.

Potential Explanatory Factors

Explanatory factors that potentially affect water quality include land use, depth (well depth and the depth to the top of the uppermost open interval), groundwater age, and geochemical conditions. Sources and methodologies for obtaining data for these factors are discussed in the following sections. Potential correlations within these factors also were evaluated to identify which factors are likely to relate directly to water quality and could result in higher relative-concentrations or detection frequencies, and which factors may be coincidental and not directly affecting water-quality.

Land Use

Land use around wells sampled in the San Diego study unit generally indicated the composition of land use in the respective study areas as a whole. This also was true of the land use around PSWs in the CDPH database that was used in this study. The majority of land use around PSWs used in this study was natural, with lesser amounts of urban and agricultural (fig. 3A–B). The most urbanized areas around PSWs was in the Alluvial Basins study area (28 percent), followed by the Temecula Valley (23 percent), and then the Hard Rock study areas (8 percent). The Warner Valley study unit did not have wells located in any urban land-use settings. Agricultural land-use around PSWs most often was in the Temecula Valley (29 percent) study area, followed by the Alluvial Basins (17 percent), Hard Rock (1 percent) and Warner Valley (1 percent) study areas.

Well Depth

Well-construction information, including well depths, depths to the tops of the uppermost open interval, and lengths of the perforated intervals, where available, is reported in table A1. Depths for the PSWs sampled in the San Diego study unit (grid and understanding) ranged from 46 to 2,500 ft, with a median of 450 ft (fig. 7). Depth to the top of the uppermost open interval ranged from 20 to 690 feet, with a median of 96 feet. The open length ranged from 23 to 1913 feet with a median of 325 feet. These values represent different sets of wells because the total well depth was not known for as many wells as depth to the top of the uppermost open interval.

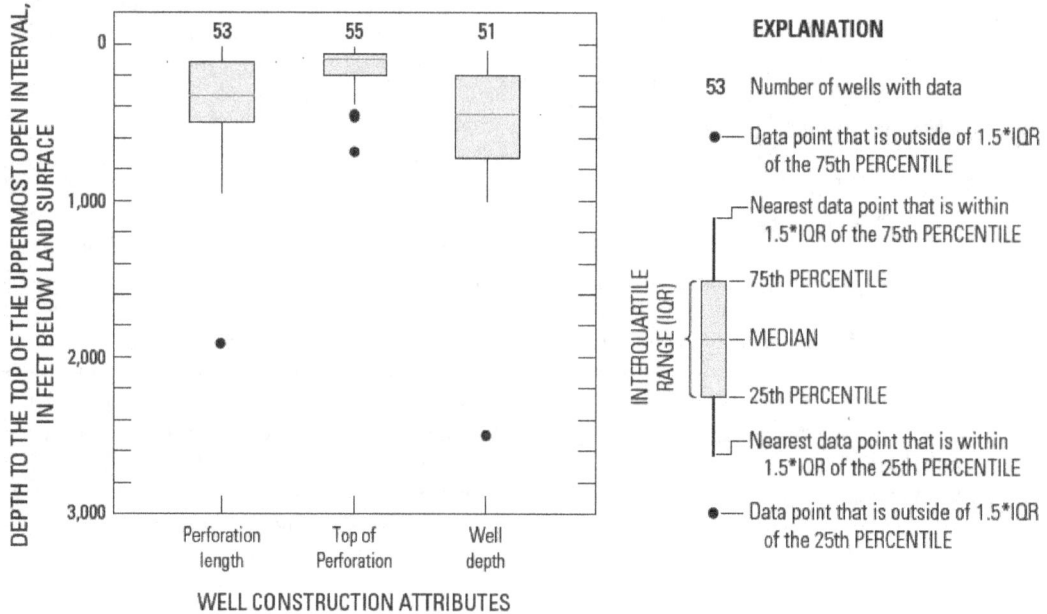

Figure 7.   Boxplots of construction attributes for grid and understanding wells, San Diego Groundwater Ambient Monitoring and Assessment (GAMA) study unit, California, May–July 2004.

Groundwater Age Classification

Of the 58 groundwater samples collected by the USGS in the San Diego study unit, 8 were modern, 29 were mixed, and 19 were pre-modern (see table C1). Samples from two wells could not be classified because the age-tracer data was incomplete or did not meet all quality-assurance checks. Classified groundwater ages generally were older with increased depth to the top of the uppermost open interval (fig. 8A). The depth to the top of uppermost open interval was significantly less for wells with modern and mixed age distributions than for wells with pre-modern age distributions. Relative to well depth, wells classified as modern and mixed were significantly shallower than wells classified as pre-modern (fig. 8B).

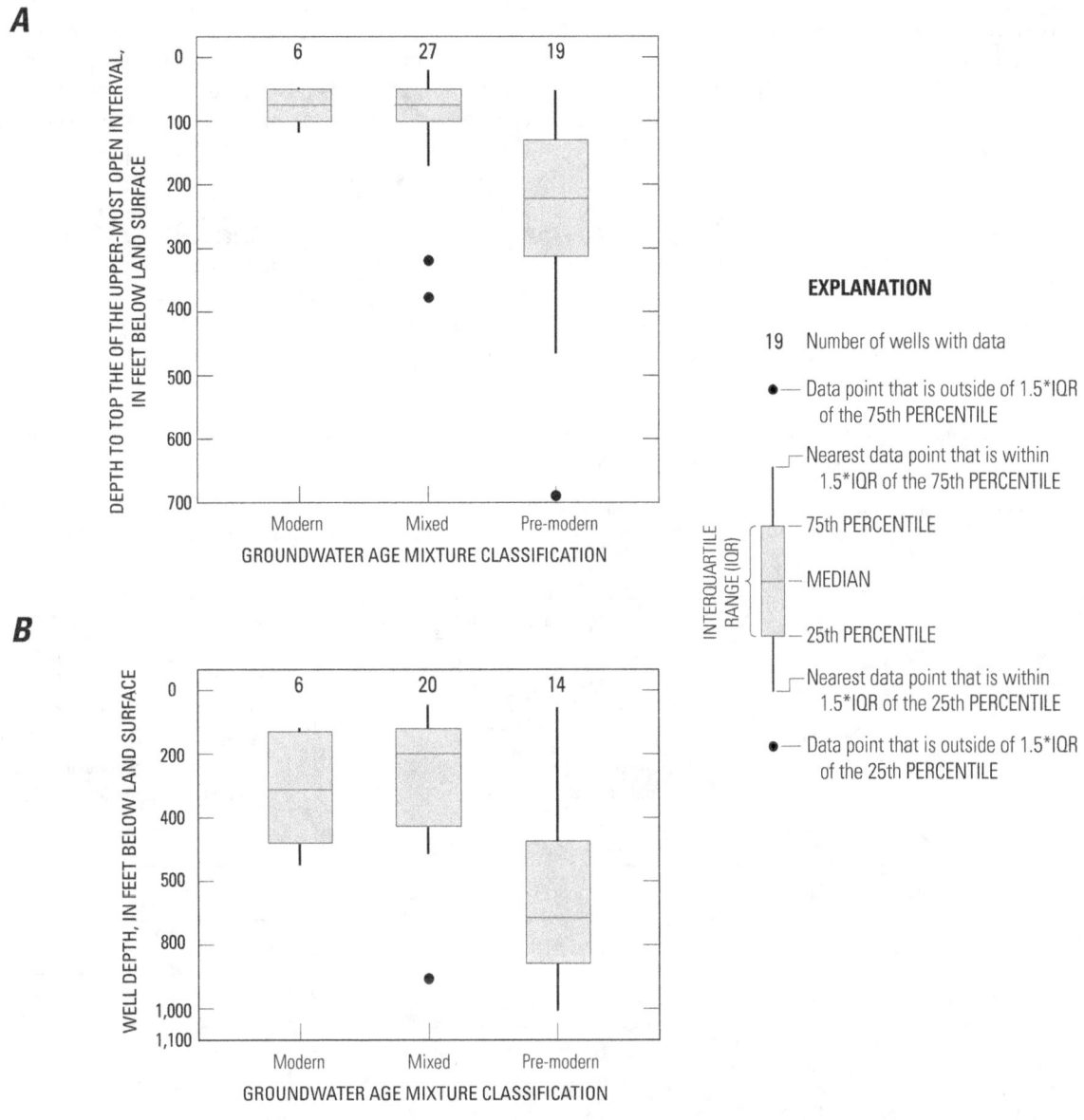

Figure 8A–B.   Boxplots of relation of classified groundwater age to (A) depth to top of the uppermost open interval below land surface and (B) well depth below land surface, San Diego Groundwater Ambient Monitoring and Assessment (GAMA) study unit, California, May–July 2004.

Geochemical Condition

Geochemical information collected for the San Diego study unit included pH, dissolved oxygen (DO), and concentrations of nitrate, manganese, and iron. Concentrations of DO, nitrate, manganese, and iron were used to determine the "redox" (oxidation-reduction) condition for the wells, using techniques described in appendix C. In the San Diego study unit, data was sufficient to classify the redox condition for 45 grid and understanding wells. Wells were either classified as oxic or anoxic; wells tapping groundwater with a mixed redox condition were not used this analysis. Sixty-two percent of the wells were classified as anoxic and 38 percent as oxic. pH values in the study unit ranged from 6.6 to 9.5 with a median value of 7.4.

Correlations between Explanatory Variables

Apparent correlations between an explanatory variable and a water-quality constituent actually could indicate correlations between explanatory factors. For example, detections of VOCs may be inversely correlated to urban land-use in a given area because the uppermost open interval of wells tend to be deep, and the water being tapped is pre-modern, not because VOCs are not used in urban settings. Therefore, it is important to identify statistically significant correlations between explanatory variables

The majority of explanatory variables used in this report are not significantly related (table 7). The strongest correlation is between well depth and depth to the top of the uppermost open interval. Because of the significance of this correlation only depth to the top of the uppermost open interval will be used in this report. Positive correlations of well depth to groundwater classified as pre-modern and pH were significant. The only other significant correlations were positive correlations between pH and groundwater classified as pre-modern and between anoxic groundwater and urban land-use; there was a negative correlation between natural land-use and depth to the top of the uppermost open interval.

Table 7.   Results of non-parametric analysis of correlations between selected potential explanatory variables, San Diego Groundwater Ambient Monitoring and Assessment (GAMA) study unit, California, May–July 2004.

[Results are shown only for those correlations with a p-value ≤ 0.1. Results with p-values ≤ 0.05 are shown in bold. Only results with p-values ≤ 0.05 are considered significant in this study. ρ, Spearman's correlation statistic; Z, test statistic for Wilcoxon test; negative number is inverse relation between variables; –, p >0.1; <, less than; $\leq$, less than or equal to]

Wells included in analysis	Explanatory factor	ρ: Spearman's correlation statistic			Z: Wilcoxon test statistic			
		Depth to top of the upper-most open interval, feet below land surface	Depth of well below land surface, feet	pH, pH units	Anoxic versus oxic	Mixed versus modern age class	Modern versus pre-modern age class	Mixed versus pre-modern age class
Grid wells	Percentage urban land use	–	–	–	**2.02**	–	–	–
	Percentage agricultural land use	0.27	–	–	–	1.40	–	–
	Percentage natural land use	**-0.39**	–	–	–1.65	–	–	–
Grid and understanding wells	Depth to the top of uppermost open interval below land surface, feet		**0.73**	–	–	–	**–3.10**	**–4.10**
	Depth of well below land surface, feet			**0.54**	–	–	**–2.93**	**–4.13**
	pH, pH units				–	–	**–2.77**	**–3.40**

Status and Understanding of Water Quality

As a starting point for summarizing the results of approximately 16,000 individual analytical measurements in the San Diego study unit, the maximum relative-concentrations of the individual constituents and constituent groups were calculated for all four study areas (fig. 9). Health-based benchmarks are established for all constituents shown, except for those in the group inorganic-SMCL, for which non-health-based aesthetic benchmarks are established. Aquifer proportions calculated by the grid-based approach were considered the most reliable and are used in the subsequent discussions, except where otherwise noted. In some instances, the spatially weighted approach identified constituents that could be present at moderate or high relative-concentrations in small proportions of the primary aquifers that were not identified using the grid-based approach. Results from the spatially weighted approach were used only in cases for which the grid-based approach was found to have this limitation. Non-significant relations generally are not discussed; selected significant correlations are shown graphically.

Thirty-four of the 218 organic and special-interest constituents analyzed for were detected in samples collected at grid wells (table 5). Some type of health-based benchmark has been established for most of the organic and special interest constituents detected (23 of the 34). Five of the constituents with no health-based benchmarks are pesticide degradates. Some of the parent compounds (atrazine, diuron) of these degradates with health-based benchmarks were detected in samples. In contrast to organic and special-interest constituents, inorganic constituents were nearly always detected (48 of 50, table 5). Health-based or aesthetic benchmarks were not established for just over one-quarter of inorganic constituents (13 of 48). Most of the constituents without benchmarks are major or minor ions that are naturally present in groundwater.

Table 4 shows the area-weighted aquifer-scale proportions for the Temecula Valley, Warner Valley and Alluvial Basins study areas (hereinafter referred to as the Alluvial Fill study areas because they are composed of alluvial fill aquifers), and tables B1A–D show aquifer-scale proportions for the individual study areas. Aquifer-scale proportions in these tables are calculated by using both the grid-based and spatially weighted methods, and show constituents with high relative-concentrations under the following criteria: (1) high relative-concentrations detected during sampling for the GAMA Priority Basin Project, (2) high relative-concentrations in the CDPH database during the current period (July 30, 2001–July 29, 2004), and (3) historically high relative-concentrations in the CDPH database.

Inorganic Constituents

Sixteen inorganic constituents qualified as constituents of interest because their relative-concentrations were greater than 0.5 in the grid-based assessment (fig. 10). Inorganic constituents with health-based benchmarks (nutrients, trace elements, and radioactive constituents) were high in 17.6 percent of the primary aquifers in the Alluvial Fill study areas (table 8). The greatest proportion of the primary aquifers with high relative-concentrations is in the Temecula Valley (27.3 percent) and Alluvial Basins (13.3 percent) study areas, whereas no high relative-concentrations were detected in the Warner Valley study area (tables B2A–C). High relative-concentrations were observed in 25.0 percent of the primary aquifers in the Hard Rock study area (table B2D).

Trace Elements

The relative-concentrations of trace elements meeting the selection criteria (relative-concentration ≥ 0.5) are shown in figure 10. Trace elements were detected at high relative-concentrations in 14.4 percent of the primary aquifers in the Alluvial Fill study areas (table 8). The greatest proportion of the primary aquifers with high relative-concentrations was in the Temecula Valley (27.3 percent) and Alluvial Basins (6.7 percent) study areas (tables B2A and C). High relative-concentrations (based on spatially weighted calculations) were detected in 1.2 of the primary aquifers of the Hard Rock study area (table B2D). The three trace elements that were detected at high relative-concentrations in greater than or equal to 2 percent of the primary aquifers (based on aquifer-scale proportion that were not area-weighted for all study areas) were vanadium (2.8 percent), arsenic (2.0 percent), and boron (2.0 percent); the distribution and factors affecting distribution of these trace elements are discussed in more detail below.

The location and distribution of V, As, and B in the San Diego study unit are displayed on figures 11A–C. Of the high relative-concentrations detected for these trace elements, only a single high detection (V) was observed outside of the Temecula Valley study area. Moderate relative-concentrations for these trace elements also were most frequently detected in the Temecula Valley study area.

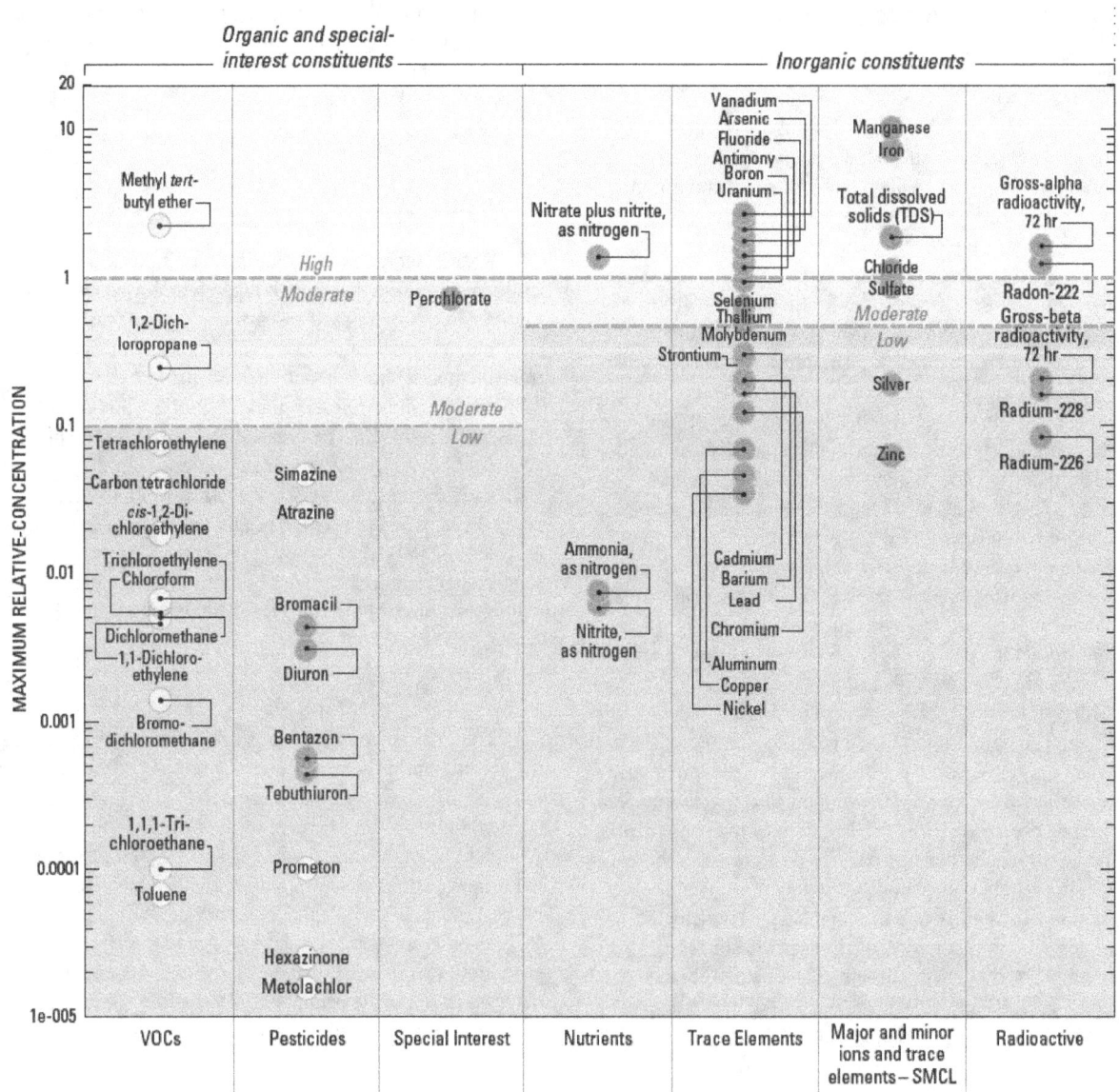

EXPLANATION

Zinc   **Constituents collected at all grid wells** — Name and center of symbol is location of data unless indicated by following location line: ⌐

Diuron   **Constituents collected at a subset of grid wells** — Name and center of symbol is location of data unless indicated by following location line: ⌐

SMCL, secondary maximum contaminant level

Figure 9.   Maximum relative-concentration in grid wells for constituents detected by type of constituent in the San Diego Groundwater Ambient Monitoring and Assessment (GAMA) study unit, California, May–July 2004.

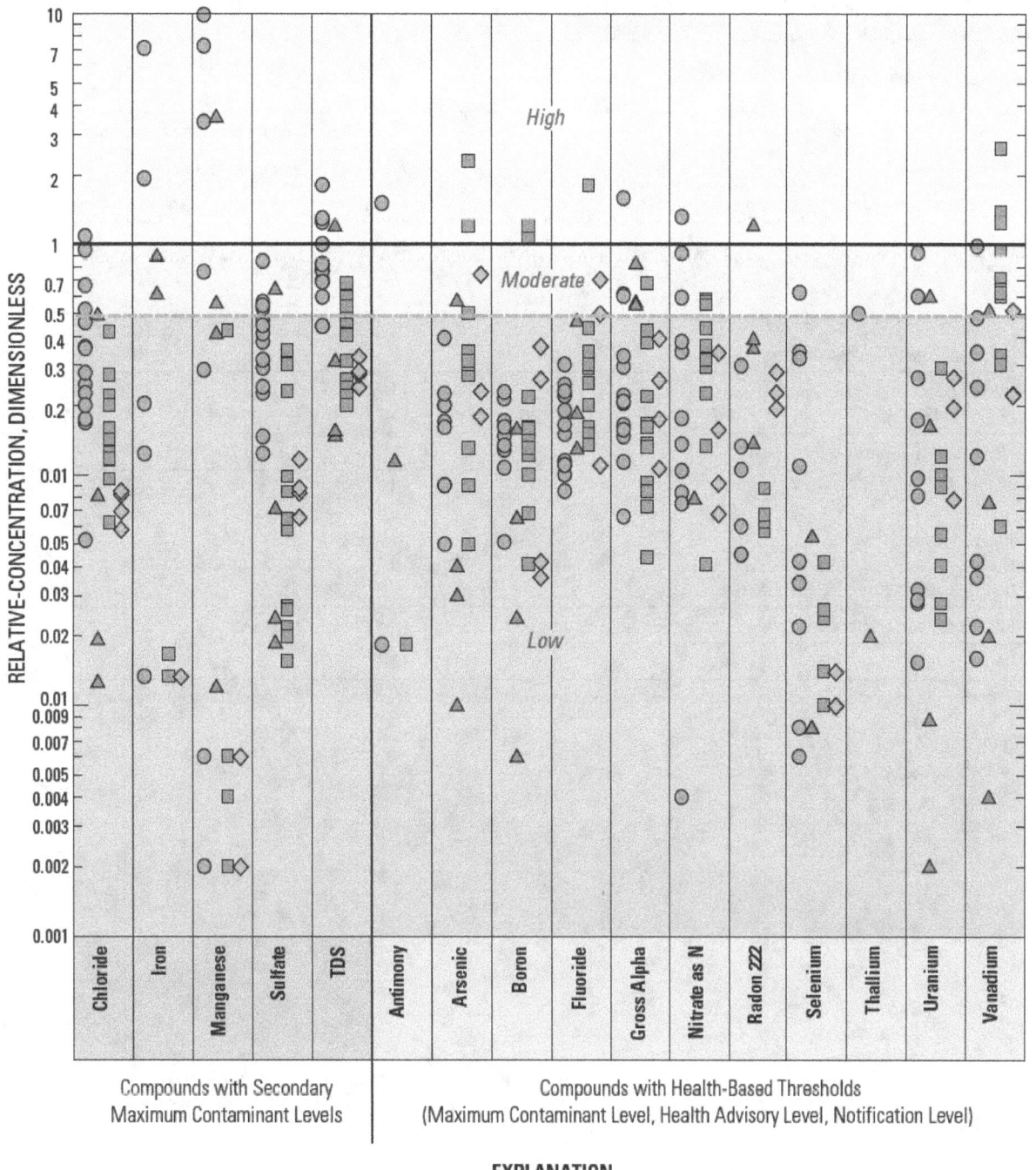

Figure 10. Dot plots of relative-concentrations of selected trace elements, radioactive constituents, nutrients, and major and minor elements in grid wells, San Diego Groundwater Ambient Monitoring and Assessment (GAMA) study unit, California, May–July 2004.

Table 8.   Grid-based aquifer-scale proportions for constituent classes in the Alluvial Fill study areas, (Temecula Valley, Warner Valley, and Alluvial Basins), San Diego Groundwater Ambient Monitoring and Assessment (GAMA) study unit, California.

[Values are grid based unless otherwise noted]

Constituent class	Aquifer-scale proportion[1] (percent)		
	High values	**Moderate values**	**Low values**
Inorganics with health-based benchmarks			
Trace elements	14.4	27.8	57.8
Radioactive	3.2	13.7	83.1
Nutrients	3.4	6.8	89.8
Any inorganic with health-based benchmarks	17.6	32.3	50.1
Inorganics with aesthetic benchmarks			
Total dissolved solids and (or) chloride and (or) sulfate	13.7	31.2	55.1
Manganese and (or) iron	13.7	3.4	82.9
Organics with health-based benchmarks			
Trihalomethanes	0.0	0.0	100.0
Solvents	0.0	3.0	97.0
Gasoline components	3.0	0.0	97.0
Pesticides	0.0	0.0	100.0
Any organic with health-based benchmarks	3.0	3.0	94.0
Constituents of special interest			
Perchlorate	[2]0.2	36.3	63.7

[1] Alluvial Fill study areas aquifer-scale proportion is calculated by summing the area-weighted average for each individual study area except the Hard Rock. Area-weighted values for each study area are: Temecula Valley = 0.41, Warner Valley = 0.11, Alluvial Basins = 0.48. Aquifer-scale proportions will not sum to 100 if a spatially weighted value is used.

[2] Spatially weighted value.

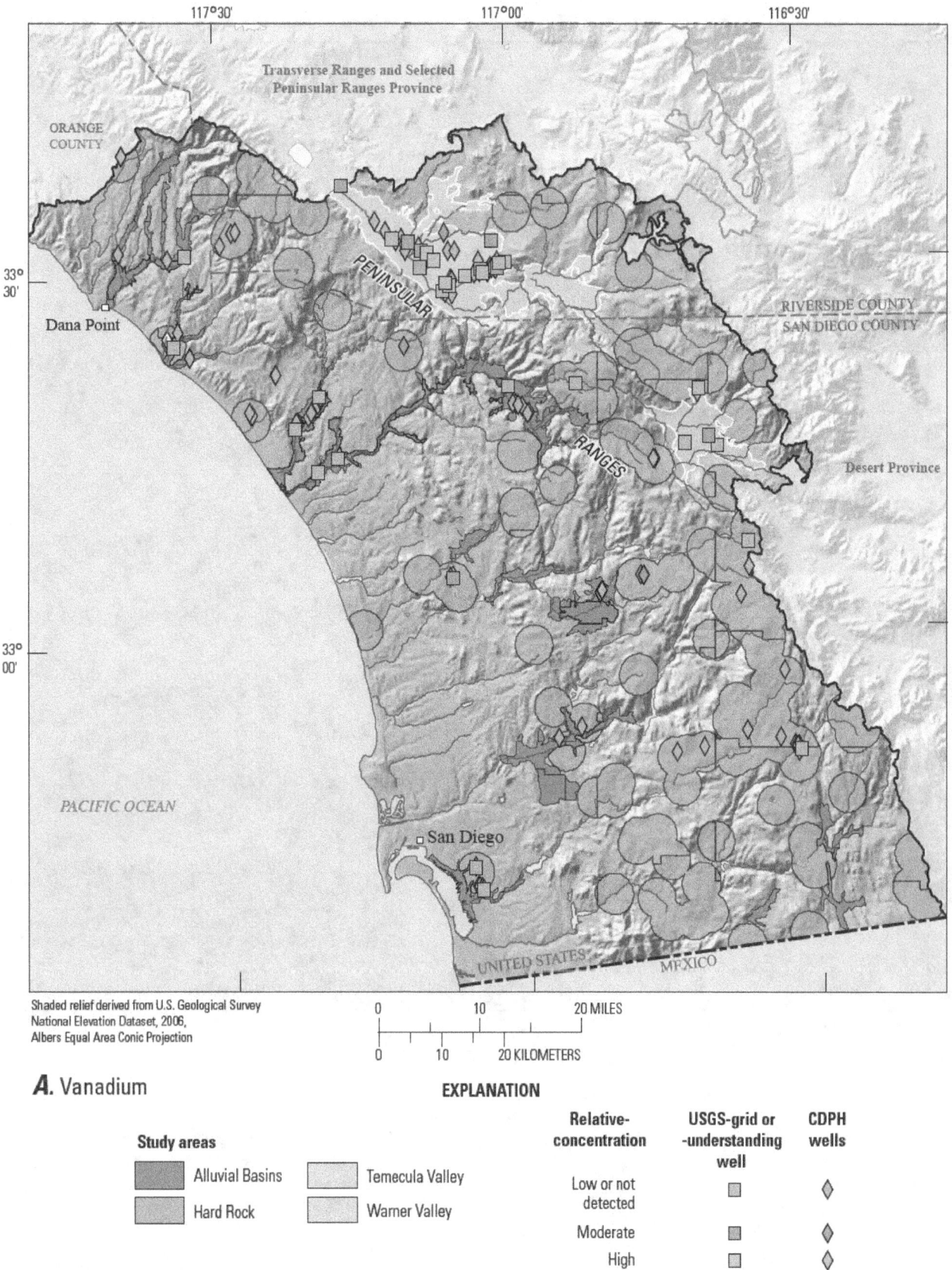

Shaded relief derived from U.S. Geological Survey
National Elevation Dataset, 2006,
Albers Equal Area Conic Projection

0   10   20 MILES

0   10   20 KILOMETERS

A. Vanadium

EXPLANATION

Study areas

Alluvial Basins

Hard Rock

Temecula Valley

Warner Valley

Relative-concentration	USGS-grid or -understanding well	CDPH wells
Low or not detected		
Moderate		
High		

Figure 11A–C.   Values of selected inorganic constituents in USGS-grid and -understanding wells representative of the primary aquifers and the most recent analysis (July 30, 2001–July 29, 2004) for CDPH wells, San Diego Groundwater Ambient Monitoring and Assessment (GAMA) study unit, California, May–July 2004: (A) vanadium, (B) arsenic, and (C) boron.

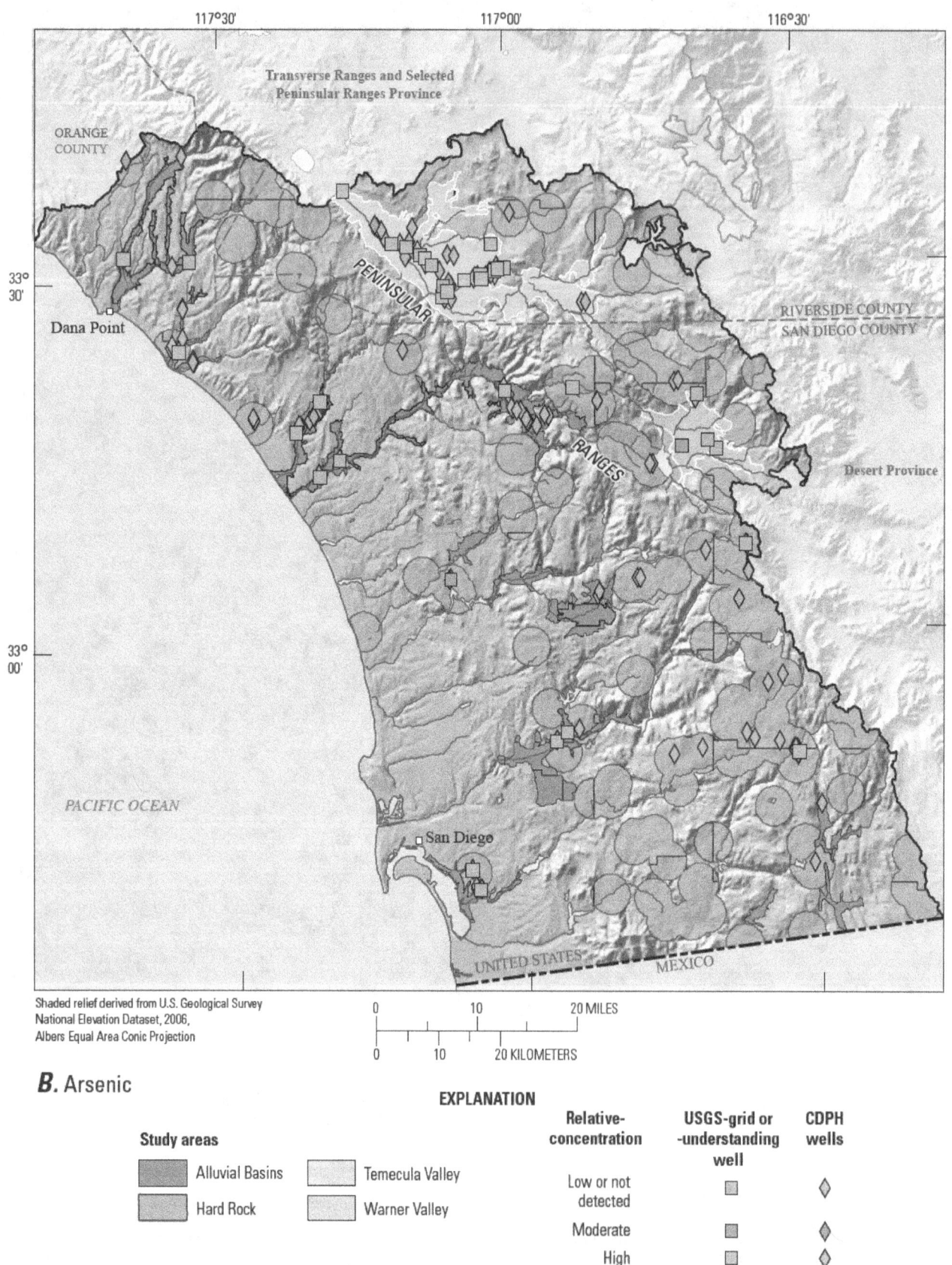

Shaded relief derived from U.S. Geological Survey
National Elevation Dataset, 2006,
Albers Equal Area Conic Projection

B. Arsenic

EXPLANATION

Study areas

▨ Alluvial Basins	☐ Temecula Valley	
▨ Hard Rock	☐ Warner Valley	

Relative-concentration	USGS-grid or -understanding well	CDPH wells
Low or not detected	▪	◇
Moderate	▪	◇
High	▪	◇

Figure 11*A–C.*—Continued

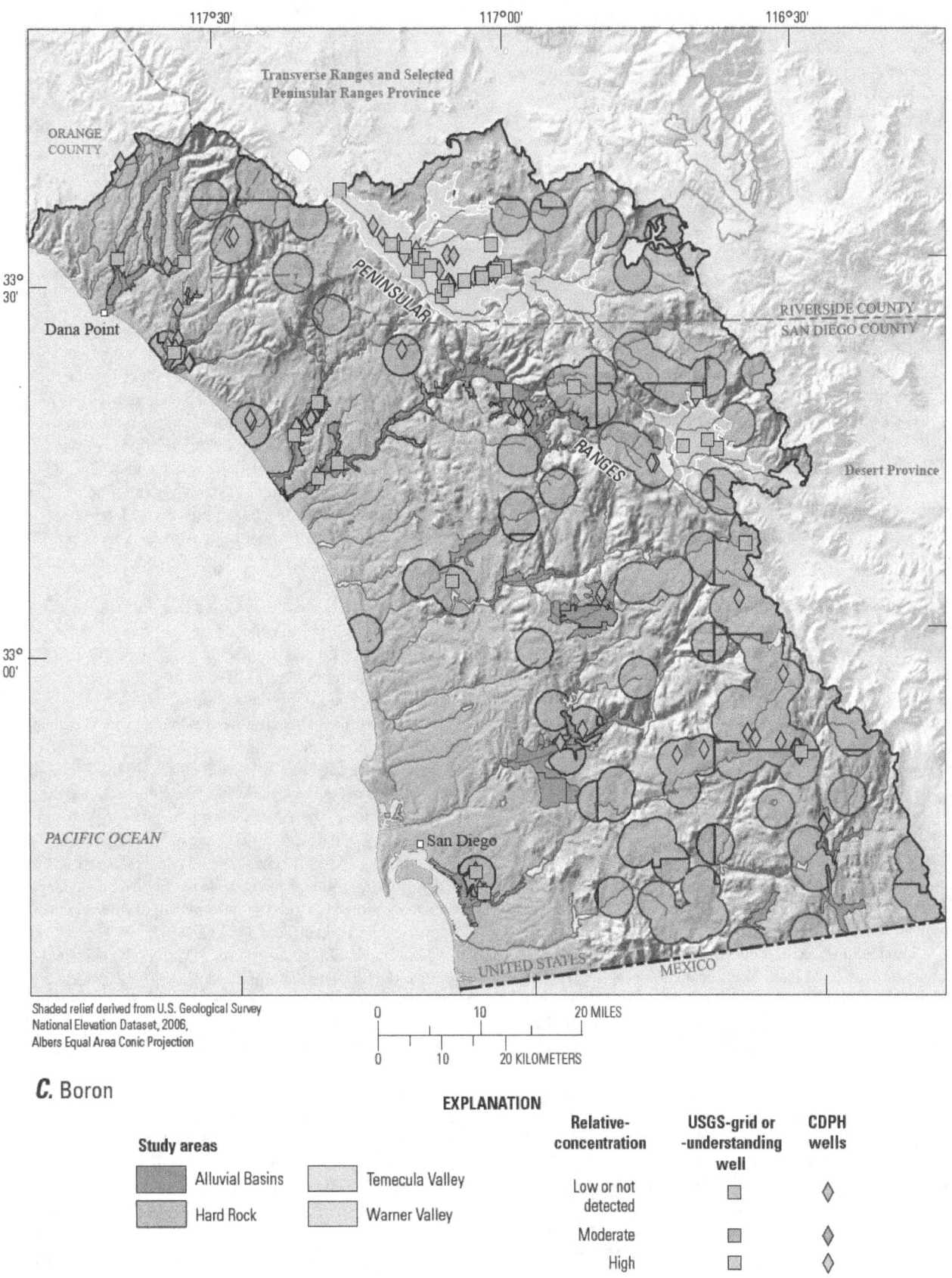

117°30'   117°00'   116°30'

Transverse Ranges and Selected
Peninsular Ranges Province

ORANGE
COUNTY

33°
30'

Dana Point

RIVERSIDE COUNTY
SAN DIEGO COUNTY

PENINSULAR

Desert Province

RANGES

33°
00'

PACIFIC OCEAN

San Diego

UNITED STATES   MEXICO

Shaded relief derived from U.S. Geological Survey
National Elevation Dataset, 2006,
Albers Equal Area Conic Projection

0   10   20 MILES

0   10   20 KILOMETERS

C. Boron

EXPLANATION

Study areas		Relative-concentration	USGS-grid or -understanding well	CDPH wells
Alluvial Basins	Temecula Valley	Low or not detected	▪	◇
Hard Rock	Warner Valley	Moderate	▪	◇
		High	▪	◇

Figure 11*A–C.*—Continued

Factors Affecting Vanadium Distribution

Potential sources of V to groundwater are both natural and anthropogenic. Natural sources can be attributed to the dissolution of V-rich rocks, which include mafic rocks such as basalts and gabbros (Nriagu, 1998), and sedimentary rocks such as shale (Vine and Tourtelet, 1970; McKelvey and others, 1986). Anthropogenic sources of V can come from waste streams associated with the ferrous metallurgy industry (World Health Organization, 1988) and through the combustion of V-enriched fossil fuels, primarily in the form of residual crude oil and coal (Duce and Hoffman, 1976; Hope, 1997). Atmospheric V can be deposited to the land surface through wet and dry deposition and transported into to the subsurface by infiltrating surface water.

The results of a previous study by Wright and Belitz (2010) indicated that the source of moderate and high relative-concentrations of V (> 25 µg/L) in California, and in particular the Temecula Valley, likely is mafic and andesitic rock. In the San Diego study unit, correlations between land use and V concentrations in samples collected for this study did not indicate that anthropogenic activities were significant contributing sources (table 9), which implies that V-rich rocks are likely the significant contributing source of V to groundwater in the San Diego study unit.

The redox conditions of the system under considerations will influence V concentrations in groundwater. This is because V is a redox sensitive element that exists in three oxidation states in the environment: V (III), V (IV), and V (V). Thermodynamically speaking, the predominant oxidation state of V is dependent on the Eh and pH conditions of the aqueous system under consideration. Vanadium (V) and V (IV) are the most important species in natural waters, with V (V) likely the most abundant under environmental conditions (Hem, 1985). The solubility of V in groundwater is likely to be largely controlled by adsorption/desorption processes on mineral surfaces (Wehrli and Stumm, 1989; Wanty and others, 1990; Wanty and Goldhaber, 1992). Vanadium (V), an oxyanion, and V (IV), an oxycation, both adsorb to mineral surfaces. However, under most environmental conditions V is expected to be most mobile under oxic and alkaline conditions.

Vanadium concentrations were significantly higher in samples collected from oxic and alkaline (high pH) groundwater than in samples collected from anoxic groundwater (fig. 12A; table 9). Vanadium was detected at high or moderate relative-concentrations only in samples collected from oxic groundwater; concentrations were less than or equal to 10 µg/L for all samples collected from anoxic groundwater. Additionally, the four samples with the highest concentrations were collected from groundwater with a pH of at least 7.9 (fig. 12A). These results indicate that V is indeed being desorbed from, or being inhibited from adsorbing to, mineral surfaces under oxic and alkaline conditions.

The highest V concentrations tended to be detected in samples collected from deep wells with mixed and pre-modern groundwater age classifications (fig. 12B; table 9). This relation most likely is a result due in part to the fact that pH values of pre-modern groundwater generally were higher than pH values of either modern or mixed waters (table 7). In addition, 73 percent of the samples with redox indicator data that were classified as pre-modern were classified as oxic. Again, these relations highlight the relation between high V concentrations and oxic and alkaline groundwater conditions.

Factors Affecting Arsenic Distribution

Like V, potential sources of As to groundwater are both natural and anthropogenic. Natural sources may be attributed to the dissolution of relatively As-rich igneous rocks like basalts and gabbros and sedimentary marine rocks, such as shale and phosphorites (Welch and others, 1988). Anthropogenic uses of As are varied, but the dominant uses in the United States are agricultural applications, wood preservation, and glass production (Welch and others, 2000). In the San Diego study unit, the positive correlation of arsenic concentrations in groundwater samples to any land-use type was not significant, which suggests that As-rich rocks are the most significant source of arsenic concentrations to groundwater.

Arsenic also is a redox sensitive element with a behavior affected by the redox and pH conditions of the groundwater system under consideration. Arsenic is stable in two oxidation states in the environment: As (III) and As (V). Over a wide pH range and oxic conditions, the oxyanion As (V) is predicted to be the predominant species, whereas under more reducing (anoxic) conditions the oxyanion As (III) likely would be the predominant species (Welch and others, 1988). Previous investigations of As in groundwater (Belitz and others, 2003; Welch and others, 2006) and literature reviews (Welch and others, 2000; Stollenwerk, 2003) have attributed elevated As in groundwater to two mechanisms: (1) the release of As from the dissolution of iron or manganese oxyhydroxides under anoxic conditions; (2) the desorption from, or inhibition of sorption to, mineral surfaces at alkaline pH.

The distribution of sample As concentrations was not significantly correlated to either redox or pH conditions of groundwater in the San Diego study unit (fig. 13A; table 9), although concentrations were correlated to pH at the 90 percent confidence level. These results suggest that different processes, or a combination thereof, are influencing As concentrations in groundwater. Release of As from iron and (or) manganese oxyhydroxides in anoxic groundwater conditions, and (or) the desorption of As from mineral surfaces under alkaline groundwater conditions may be influencing As concentrations detected in groundwater in the San Diego study unit. Even though the statistical correlation was not significant, sample concentrations generally did increase with increasing pH, indicating that As is more available in alkaline groundwater.

Table 9. Results of non-parametric analysis of correlations between selected water-quality constituents and potential explanatory factors, San Diego Groundwater Ambient Monitoring and Assessment (GAMA) study unit, California.

[Results are shown only for those correlations with a ρ-value ≤ 0.1. Results with ρ-values ≤ 0.05 are shown in bold. Only results with ρ-values ≤ 0.05 are considered significant in this study. A positive value indicates positive correlations; negative values indicate negative correlations. ρ, Spearman's correlation statistic; Z, test statistic for Wilcoxon test; –, ρ-value > 0.1; THMs, trihalomethanes; TDS, total dissolved solids; MCL-US, U.S. Environmental Protection Agency maximum contaminant level; NL-CA, CDPH notification level; Proposed AMCL-US, U.S. Environmental Protection Agency alternative maximum contaminant level; SMCL-CA, CDPH secondary maximum contaminant level; na, not applicable; <, less than; ≤, less than or equal to; >, greater than; ≥, greater than or equal to]

Constituent	Benchmark type	High aquifer proportion (percent)	ρ: Spearman's correlation statistic	Z, Wilcoxon test statistic					ρ: Spearman's correlation statistic		
			Data analyzed: Grid and understanding wells						Data analyzed: Grid wells		
			Depth to the top of the upper-most open interval	Mixed versus modern age class	Modern versus pre-modern age class	Mixed versus pre-modern age class	Anoxic versus oxic	pH	Percent urban land use[1]	Percent agricultural land use[1]	Percent natural land use[1]
Inorganic constituents[2]											
Vanadium	NL-CA	2.8	0.33	**2.10**	–	–	**-2.50**	**0.39**	–	**-0.53**	–
Arsenic	MCL-US	2.0	**0.36**	**2.57**	–	–	–	**0.40**	–	–	–
Boron	NL-CA	2.0	–	1.72	–	–	–	**0.34**	**0.35**	–	–
Manganese	SMCL-CA	20.8	–	–	–	–	na	**-0.28**	–	–	–
TDS	SMCL-CA	17.2	**-0.31**	–	1.73	**4.10**	–	**-0.17**	–	**0.26**	–
Iron	SMCL-CA	2.0	–	–	–	–	na	**-0.38**	–	–	–
Radon-222	Proposed AMCL-US	5.3	–	–	–	–	–	–	–	–	–
Organic constituents and constituent classes[3]											
Trihalomethanes, sum of concentrations	MCL-US	0.0	–	**-1.99**	1.94	–	–	–	**0.40**	–	-0.26
Solvents, sum of concentrations	variable	0.0	–	–	–	–	–	–	**0.40**	–	**-0.30**
Gasoline, sum of concentrations	MCL-US	0.1	–	–	–	–	–	–	–	–	–
Pesticides, sum of concentrations	variable	0.0	0.23	–	2.71	**2.64**	**-2.60**	**-0.40**	–	–	–
Constituents of special interest[3]											
Perchlorate	MCL-US	0.0	–	–	–	1.70	–	–	–	0.30	**-0.36**

[1]Land-use percentages are within a circle with a radius of 500 meters around each well included in analysis.

[2]Constituents with ≥ 2 percent high aquifer-scale proportions based on non-area-weighted calculations for all study areas in the San Diego Groundwater Ambient Monitoring and Assessment (GAMA) study unit.

[3]Classes of compounds that include constituents with high and (or) moderate values or detection frequencies at any concentration ≥ 10 percent.

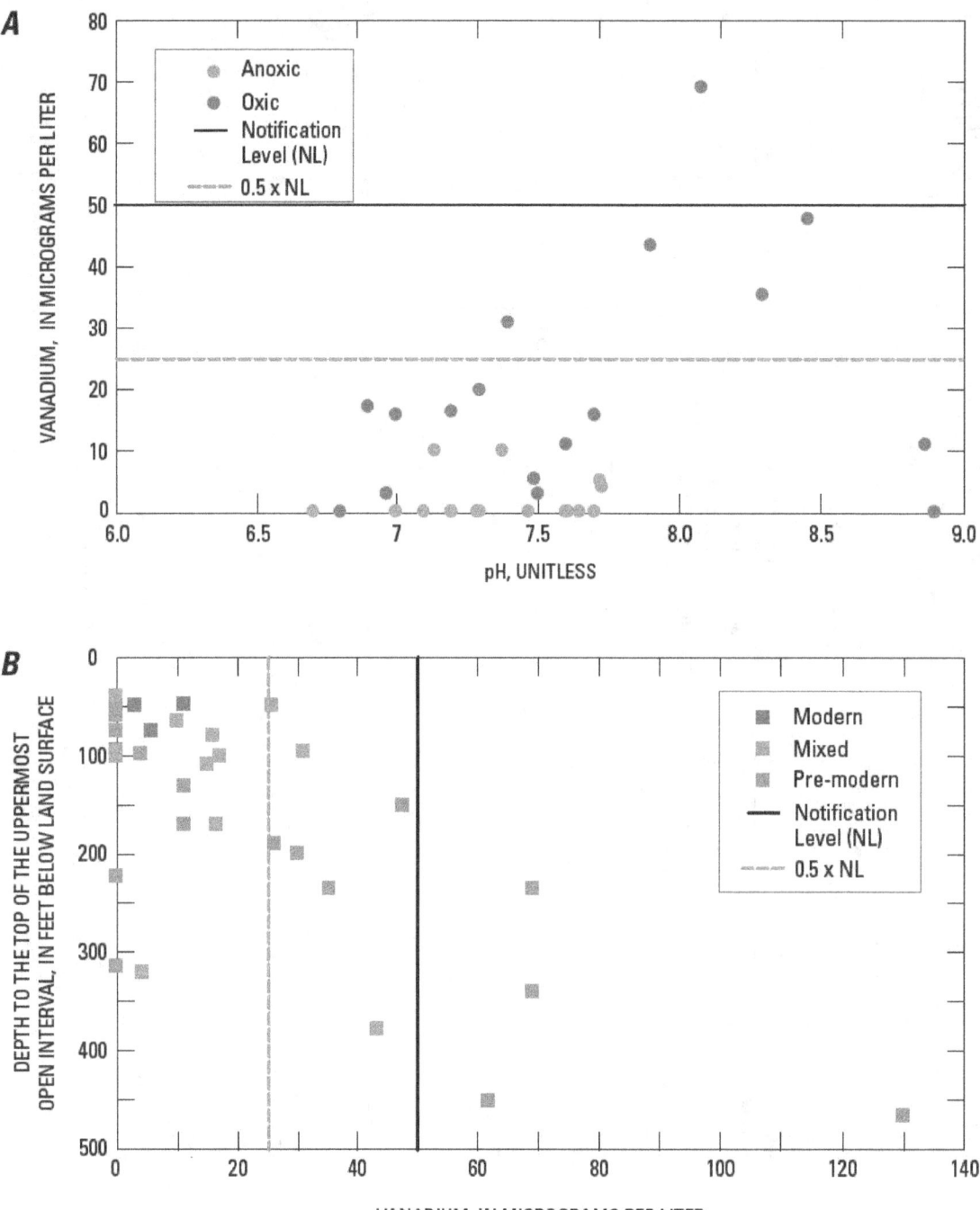

Figure 12A–B.   Relation of vanadium to explanatory variables, San Diego Groundwater Ambient Monitoring and Assessment (GAMA) study unit, California. (*A*) Relation of vanadium to redox conditions and pH and (*B*) relation of vanadium concentration to top of the uppermost open interval and groundwater age classification.

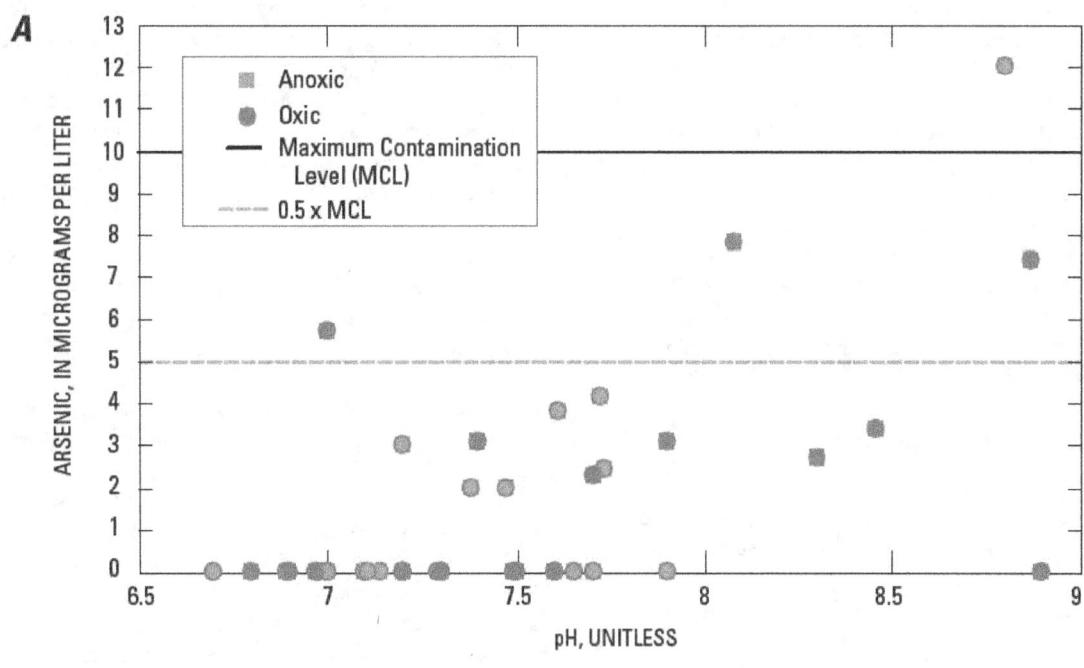

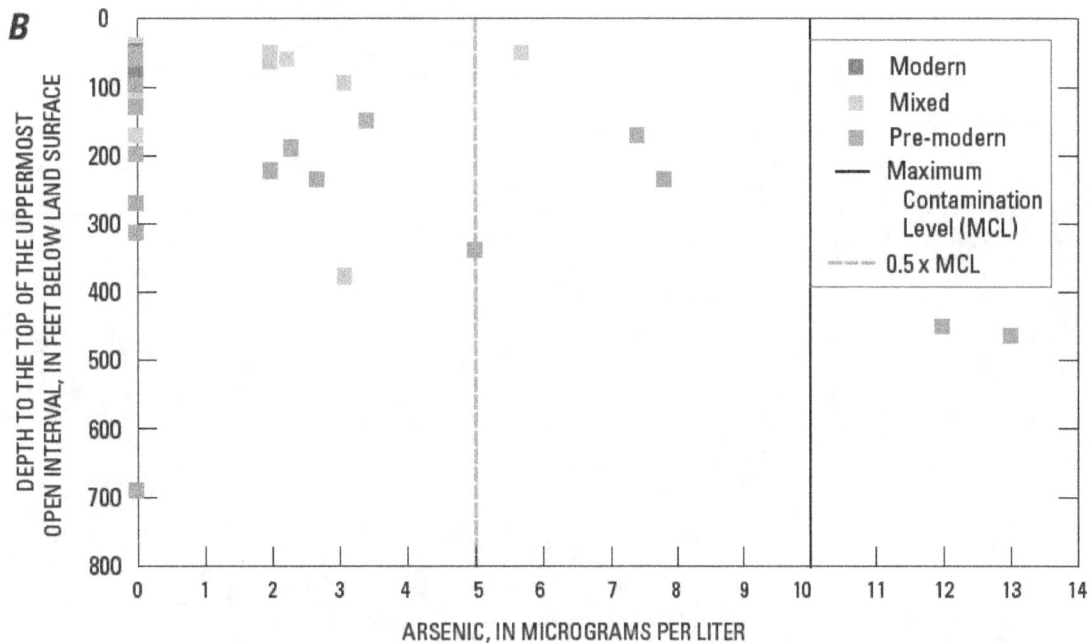

Figure 13A–B.   Relation of arsenic to explanatory variables in the San Diego Groundwater Ambient Monitoring and Assessment (GAMA) study unit, California, May–July 2004. (A) Relation of arsenic to redox conditions and pH and (B) relation of arsenic concentration to depth to the top of the uppermost open interval and groundwater age classification.

Correlations of samples with the highest As concentrations to depth to the top of the uppermost open interval and to mixed rather than to modern aged groundwater were significant (fig. 13B; table 9). Although the statistical correlation between pre-modern water and As concentrations was not significant, 83 percent of the samples with moderate and high concentrations (≥ 5 µg/L) came from samples consisting of pre-modern groundwater. The reason As concentrations tend to be highest in deep wells that are tapping mixed and pre-modern groundwater likely is a result in part that older groundwater tends to have an alkaline pH. The median pH values for samples classified as modern, mixed, and pre-modern were 7.0, 7.2, and 8.3, respectively.

Factors Affecting Boron Distribution

Natural sources of B concentrations in groundwater include the dissolution of igneous rocks like granite and pegmatites, and evaporite minerals such as kernite and colemanite (Hem, 1985; Reimann and Caritat, 1998). Borax, a B-containing evaporate mineral, is used as a cleaning agent and therefore may be present in sewage and industrial wastes. In the San Diego study unit, there was a positive correlation of B concentrations to urban land-use (table 9), indicating that

anthropogenic activities may be a source of B in groundwater. Background B concentrations are higher in seawater than in freshwater (World Health Organization 1998); therefore seawater intrusion in coastal aquifers also may increase B concentrations. Seawater intrusion does not seem to be a significant source of B in this study however, because of the relatively low concentrations of B in samples collected from the coastal alluvial aquifers (fig 11C).

Unlike V and As, B is not a redox sensitive element, and thus is not greatly affected by the redox conditions of groundwater. The molecular configuration of B in groundwater is dependent on pH, salinity, and specific cation content (Dotsika and others, 2006). The uncharged form of B, $B(OH)_3$, is predominant at pH less than 9.2, whereas the anionic form, $B(OH)_4^-$, is predominant at pH greater than 9.2. Most solid phases of B, for which data is available, are fairly soluble which suggests that adsorption and desorption reactions largely control the distribution of B in groundwater systems. In the San Diego study unit, the positive correlation between B concentrations and pH was significant (fig. 14; table 9), indicating that B is being desorbed, or inhibited from being adsorbed, to mineral surfaces under alkaline conditions. The correlations between boron and any other explanatory variables were not significant.

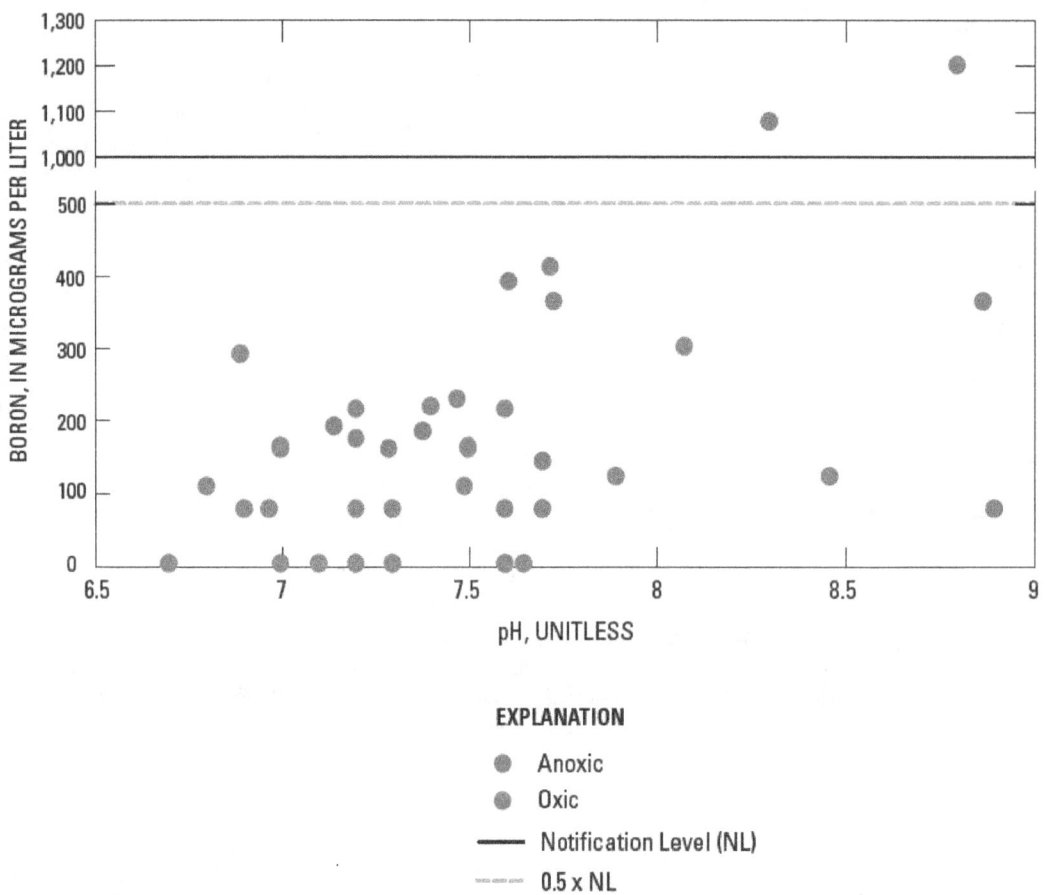

EXPLANATION

- Anoxic
- Oxic
- Notification Level (NL)
- 0.5 x NL

Figure 14. Relation of boron concentrations to redox conditions and pH, San Diego Groundwater Ambient Monitoring and Assessment (GAMA) study unit, California, May–July 2004.

Major and Minor Ions

Concentrations of some inorganic constituents can affect aesthetic properties of water, such as taste, color, and odor, and technical properties, such as scaling and staining. Although no adverse health effects are associated with these properties, consumer satisfaction with the water may be reduced or economic effects may result. For some constituents, CDPH has established non-enforceable benchmarks (SMCL-CAs) that are based on aesthetic or technical properties rather than on health-based concerns. For total dissolved solids (TDS) and the major ions chloride and sulfate, CDPH defines a "recommended" and an "upper" SMCL-CA. In this report, the "upper" SMCL-CA benchmarks were used to compute relative-concentrations. An SMCL-CA also has been established for the minor elements manganese and iron.

In the Alluvial Fill study areas, relative-concentrations of Mn and TDS were high in 13.7 percent of the primary aquifers, and relative-concentrations of Fe and fluoride (based on spatially weighted calculations) were high in 6.9 and 0.7 percent, respectively, of the primary aquifers (table 4). Manganese, TDS, and Fe were detected at high relative-concentrations in the Alluvial Basins study area at 28.6, 28.6, and 14.3 percent, respectively, and fluoride (F) was detected at high relative-concentrations (spatially weighted) in the Temecula Valley study area in 1.7 percent of the primary aquifers; major and minor ions were not detected at high relative-concentrations in the Warner Valley study area (tables B1A–C). In the Hard Rock study area Mn and TDS were detected at high relative-concentration in 33.3 and 16.7 percent of the primary aquifers, respectively, and F was detected at high relative-concentrations (spatially weighted) in 2.2 percent of the primary aquifers. Manganese (20.8 percent), TDS (17.2 percent), and Fe (2.0 percent) were the only constituents with an aesthetic benchmark that were detected at high relative-concentrations in greater than or equal to 2.0 percent of the primary aquifers for all study areas in the San Diego study unit (non area-weighted aquifer-scale proportions).

High and moderate relative-concentrations of both Mn and Fe generally occurred in the same areas of the San Diego study unit. The similar distribution of these constituents is a result of the similarities in potential sources and geochemical behavior in groundwater. High relative-concentrations of Mn and Fe were detected in every study area except for the Warner Valley (fig. 15A and 15B). High and moderate relative-concentrations most frequently were detected in the Alluvial Basins study area followed by the Hard Rock study area. In the Alluvial Basins study area, high and moderate relative-concentrations were most frequently detected in the coastal areas, whereas in the Hard Rock study area relative-concentrations were frequently highest in the most inland portions of the study area.

High relative-concentrations of TDS were detected in every study area except for the Warner Valley (fig. 15C). High relative-concentrations were most frequently detected in the Alluvial Basins study area (28.6 percent), followed by the Hard Rock study area (16.7 percent). TDS concentrations tended to be highest in the coastal and inland coastal areas of the study unit, and lowest in the most interior portions of the study unit.

Factors Affecting Manganese and Iron

Potential natural sources of Mn and Fe to groundwater include the dissolution of igneous and metamorphic rocks as well as dissolution of various secondary minerals (Hem, 1985). Rocks that contain significant amounts of Mn and Fe have a high composition of the minerals olivine, pyroxene, and amphibole. Potential anthropogenic sources of these constituents to groundwater include effluents associated with the steel and mining industries (Reimann and deCaritat, 1998). Manganese and Fe concentrations were not significantly correlated to either urban or agricultural land use (table 9), thus suggesting that natural sources are the significant contributing factor of Mn and Fe to groundwater in the San Diego study unit.

Redox and pH conditions significantly influence the concentrations of Mn and Fe in groundwater. In sediments, the oxyhydroxides of Mn and Fe are common as suspended particles and as coatings on mineral surfaces (Sparks, 1995). These oxyhydroxides are stable in oxygenated systems at neutral pH. However, under anoxic conditions, the process of reductive dissolution destabilizes these minerals which affect the mobility of Mn and Fe in aquifer systems (Sparks, 1995). Figure 16 shows the relation between DO concentrations/pH and Mn and Fe concentrations of samples collected in the San Diego study unit. The negative correlation (Spearman's rho) of both constituents to DO (Mn, rho = -0.52; Fe, rho = -0.57) and pH (table 9) was significant, indicating that reductive dissolution is a significant pathway for the mobilization of Mn and Fe in groundwater in the San Diego study unit. Manganese and Fe concentrations were not significantly correlated with any other explanatory factorss.

Factors Affecting Total Dissolved Solids

Total dissolved solids either were measured directly or calculated from specific conductance (see appendix E). Potential anthropogenic sources of TDS to groundwater in the San Diego study unit include agricultural and urban irrigation, disposal of waste water and industrial effluent, and leaking water and sewer pipes. The positive correlation of total dissolved solid concentrations to agricultural land-use in the San Diego study unit was significant (fig. 17; table 9), suggesting that agricultural irrigation practices are a significant contributing factor of TDS concentrations in groundwater.

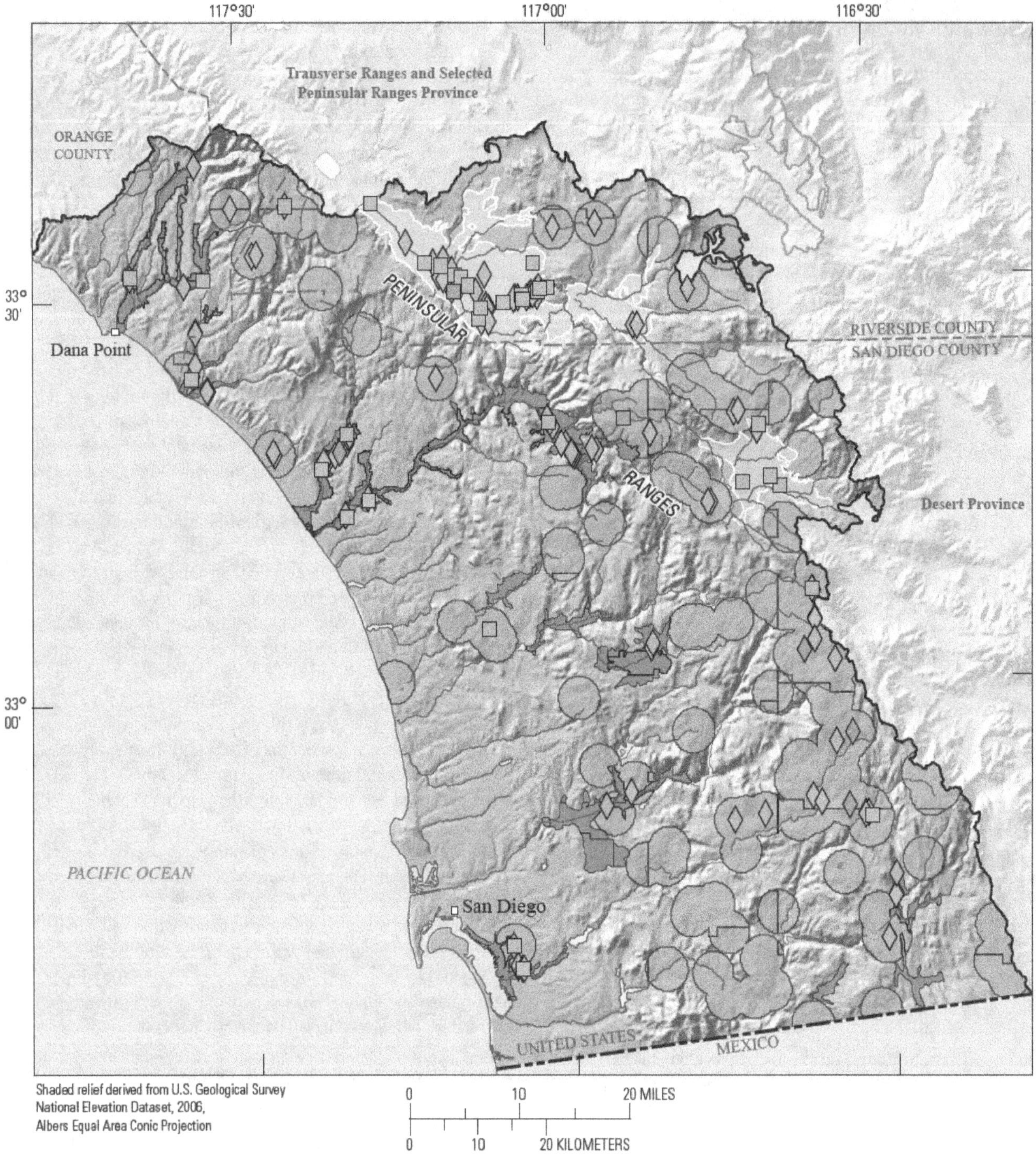

A. Manganese

EXPLANATION

Study areas

Alluvial Basins

Hard Rock

Temecula Valley

Warner Valley

Relative-concentration	USGS-grid or -understanding well	CDPH wells
Low or not detected		
Moderate		
High		

Shaded relief derived from U.S. Geological Survey
National Elevation Dataset, 2006,
Albers Equal Area Conic Projection

Figure 15A–C.   Values of selected inorganic constituents in USGS-grid and -understanding wells representative of the primary aquifers and the most recent analysis July 30, 2001–July 29, 2004, for CDPH wells, San Diego Groundwater Ambient Monitoring and Assessment (GAMA) study unit, California, May–July 2004: (*A*) manganese, (*B*) iron, and (*C*) total dissolved solids.

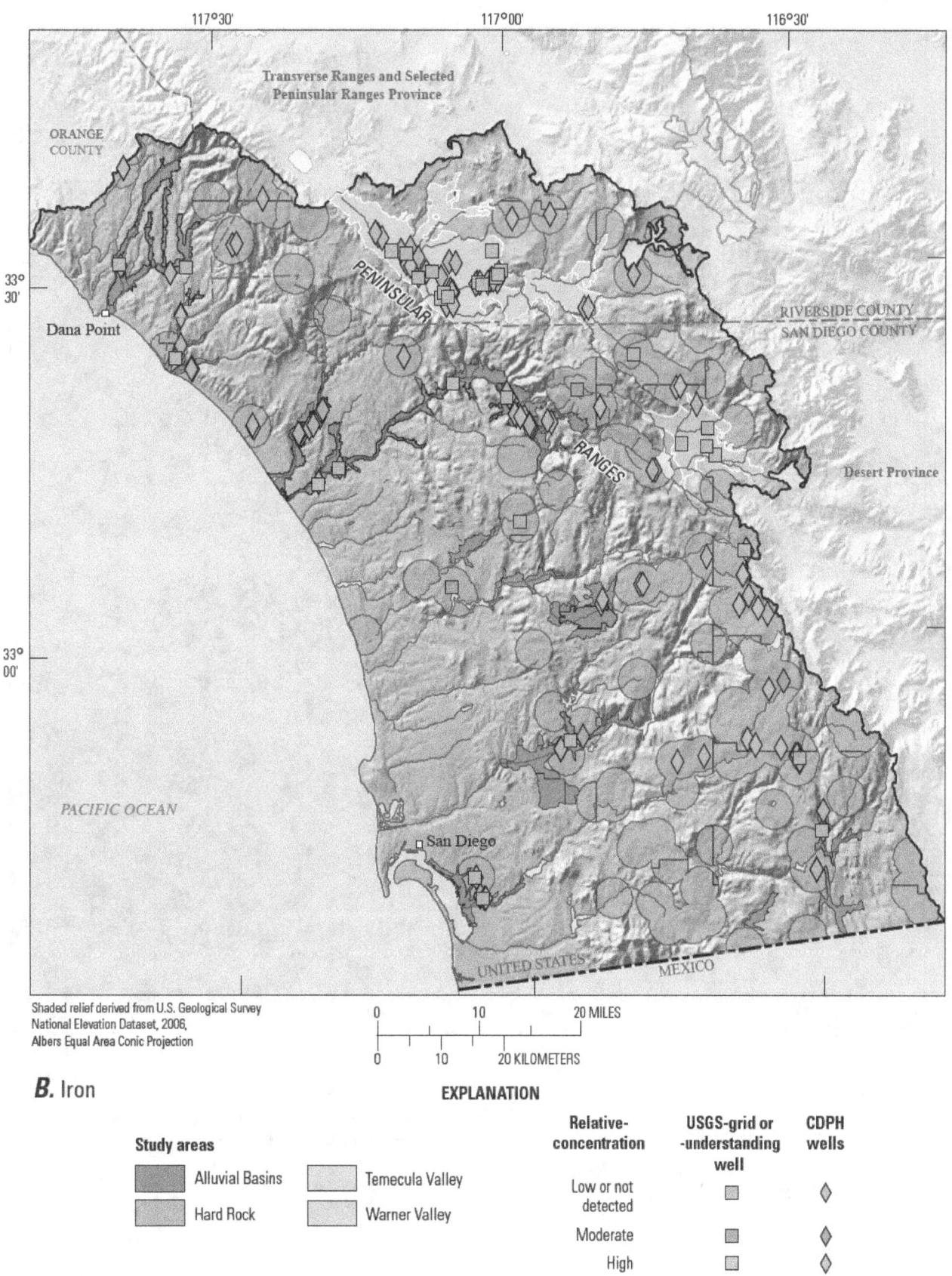

117°30'   117°00'   116°30'

Transverse Ranges and Selected
Peninsular Ranges Province

ORANGE
COUNTY

33°
30'

Dana Point

PENINSULAR

RIVERSIDE COUNTY
SAN DIEGO COUNTY

RANGES

Desert Province

33°
00'

PACIFIC OCEAN

San Diego

UNITED STATES   MEXICO

Shaded relief derived from U.S. Geological Survey
National Elevation Dataset, 2006,
Albers Equal Area Conic Projection

0   10   20 MILES

0   10   20 KILOMETERS

B. Iron   EXPLANATION

		Relative-concentration	USGS-grid or -understanding well	CDPH wells
Study areas				
Alluvial Basins	Temecula Valley	Low or not detected	▪	◇
Hard Rock	Warner Valley	Moderate	▪	◆
		High	▫	◇

Figure 15*A–C.*—Continued

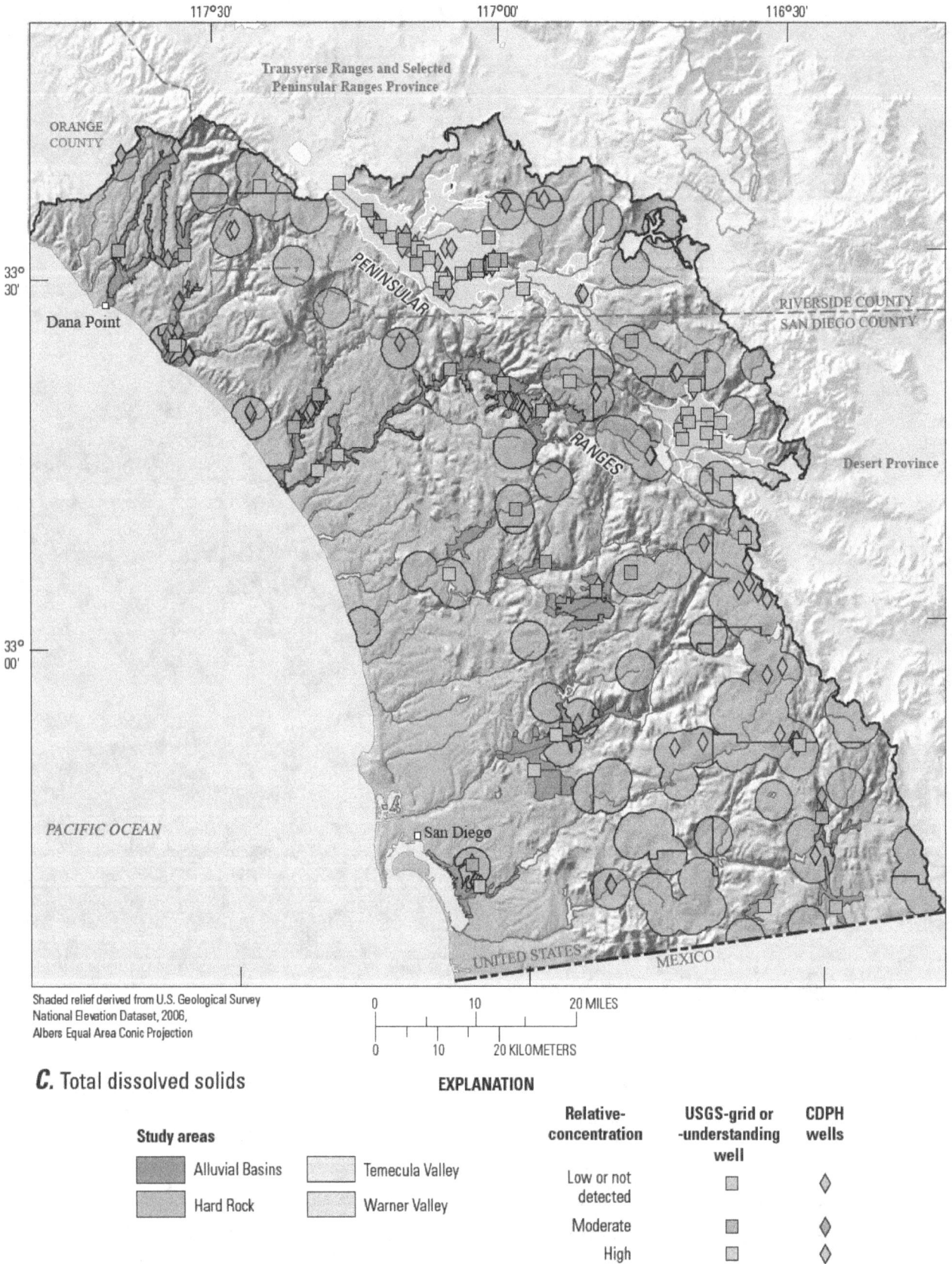

Shaded relief derived from U.S. Geological Survey
National Elevation Dataset, 2006,
Albers Equal Area Conic Projection

C. Total dissolved solids

EXPLANATION

Study areas

Alluvial Basins

Hard Rock

Temecula Valley

Warner Valley

Relative-concentration	USGS-grid or -understanding well	CDPH wells
Low or not detected		
Moderate		
High		

Figure 15*A–C.*—Continued

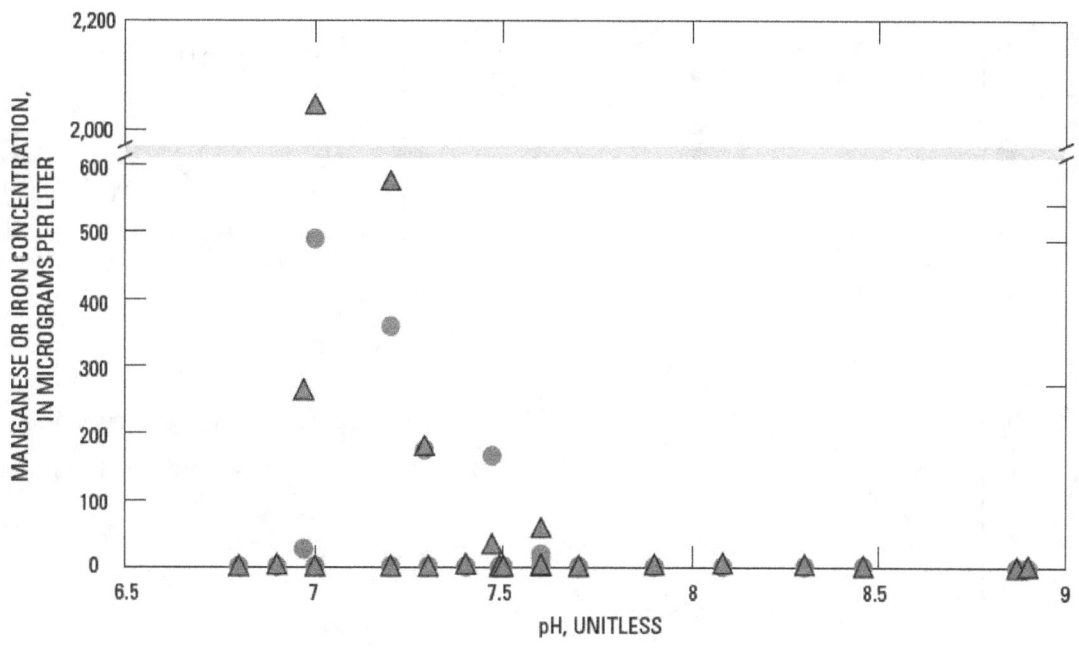

Figure 16. Relation of manganese and iron concentrations to redox conditions and pH, San Diego Groundwater Ambient Monitoring and Assessment (GAMA) study unit, California, May–July 2004.

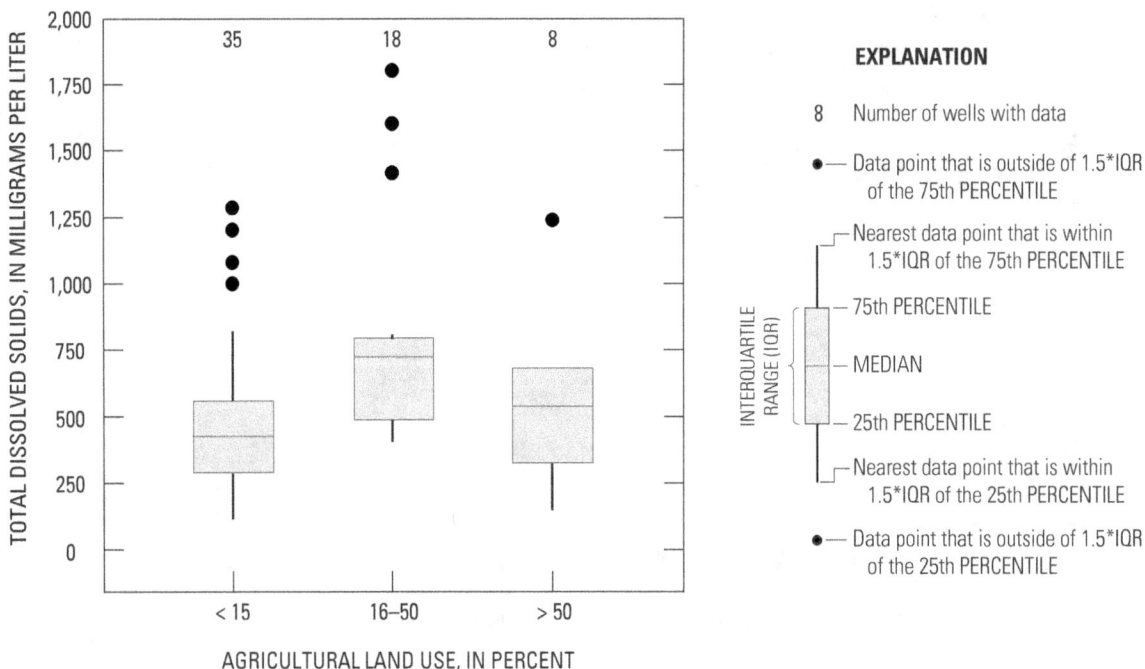

Figure 17.   Boxplots of relation of total dissolved solids (TDS) concentrations to agricultural land use, San Diego Groundwater Ambient Monitoring and Assessment (GAMA) study unit, California, May–July 2004.

Position on the flow path also may affect TDS concentrations in groundwater. Samples collected from wells at high elevations may be tapping groundwater that is located at the proximal end of flow paths where dissolution reactions with the aquifer matrix have occurred to a lesser extent than in groundwater located at the distal ends of the flow paths. Additionally, as groundwater moves down the flow path towards discharge areas, evaporation of groundwater near the water table can increase TDS concentrations.

In the San Diego study unit, the negative correlation of TDS concentrations to depth to the top of the uppermost open interval was significant as was the positive correlation of TDS to groundwater with a component of modern recharge (fig. 18; table 9). The high TDS concentrations associated with shallow wells and modern groundwater recharge implies greater loading of dissolved constituents to groundwater in

recent decades which could indicate several anthropogenic factors including agricultural and urban irrigation practices, and changes in soil chemistry as a result of historical changes in land use. The negative correlation of TDS concentrations to pH also was significant (table 9) and most likely is not the result of any geochemical processes but instead is the result of the correlation between well depth and pH (table 7).

Seawater intrusion also can cause TDS concentrations in coastal areas to increase. However, in the coastal alluvial aquifers of the San Diego study unit, previous and current studies have indicated that seawater intrusion is not a significant contributing factor to TDS concentrations (Izbicki, 1985; San Diego County Water Authority, 1997; Danskin and Church, 2005; Robert Anders, U.S. Geological Survey, personal commun., 2011).

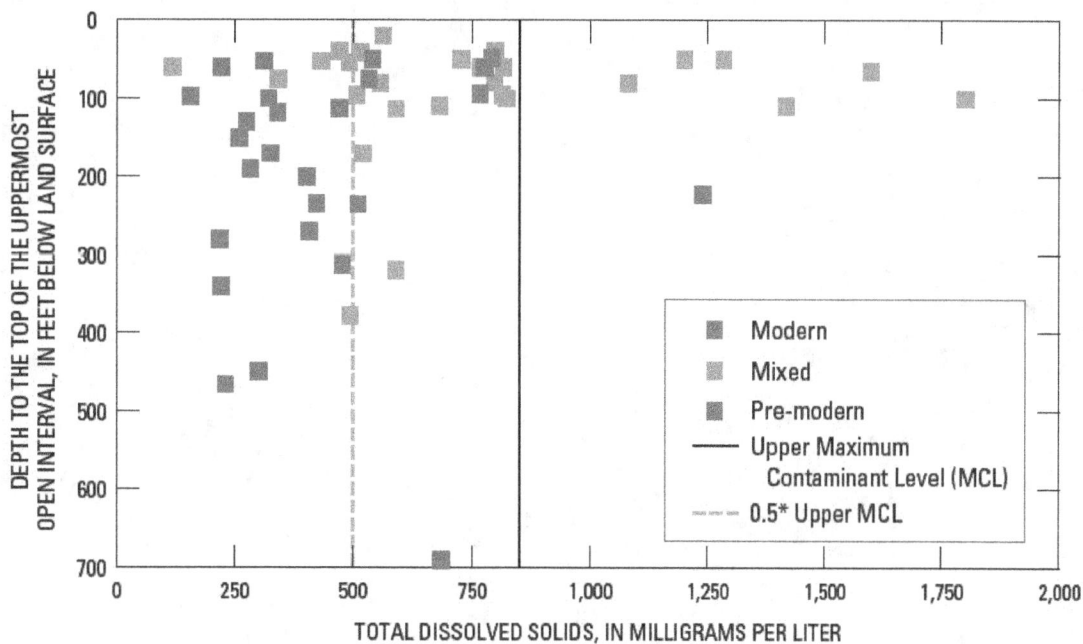

Figure 18.   Relation of TDS to depth to the top of the uppermost open interval and to groundwater age classification, San Diego Groundwater Ambient Monitoring and Assessment (GAMA) study unit, California, May–July 2004.

Radioactive Constituents

The relative-concentrations of individual radioactive constituents meeting the selection criteria (relative-concentration ≥ 0.5) are shown in figure 10. As a class, radioactive constituents were detected at high relative-concentrations in 3.2 percent of the primary aquifers in the Alluvial Fill study areas (table 8). These constituents most frequently were detected at high relative-concentrations in the Alluvial Basins (6.7 percent) study area, and were not detected at high relative-concentrations either in the Temecula Valley or in the Warner Valley study areas (tables B2A–C). In the Hard Rock study area, radioactive constituents were detected at high relative-concentrations in 25.0 percent of the primary aquifers (table B2D). Radon-222 (5.3 percent) was the only radioactive constituent detected at high relative-concentrations in greater than or equal to 2.0 percent of the primary aquifers for all study areas in the San Diego study unit (non area-weighted aquifer-scale proportions).

The location and distribution of radon-222 in the San Diego study unit is displayed in figure 19. All but one of the radon-222 detections were at low relative-concentrations. The one detection at a high relative-concentration was in the Hard Rock study area.

Factors Affecting Radon-222

Radon-222 is a radioactive gas that occurs naturally in groundwater in the decay of uranium-238 to lead-206. Uranium-238 decays in multiples steps to radium-226, which decays to radon-222 in aquifer materials. In the San Diego study unit, the correlation between radon-222 and radium-226 activities in groundwater is not significant. This insignificant correlation in part may be a result of groundwater in crystalline rocks that generally has low radium activities because radium sorbs strongly to mineral surfaces, particularly to altered feldspars (Zapecza and Szabo, 1988; Thomas and others, 1993). Radon, however, is an inert gas that readily diffuses out of the aquifer materials and into the groundwater. Ayotte and others (2007) detected greater activities of radon-222 in groundwater from crystalline bedrock aquifers in the northern United States than in aquifers comprised of glacial sediments derived from the crystalline bedrock. The greater radon-222 activities in the crystalline bedrock aquifers was attributed to concentration of sorbed radium on fracture surfaces. The highest activities of radon-222 in this study also were detected in the crystalline rock aquifers of the Hard Rock study area. Radon-222 concentrations were not significantly correlated to any of the potential explanatory variables listed in table 9.

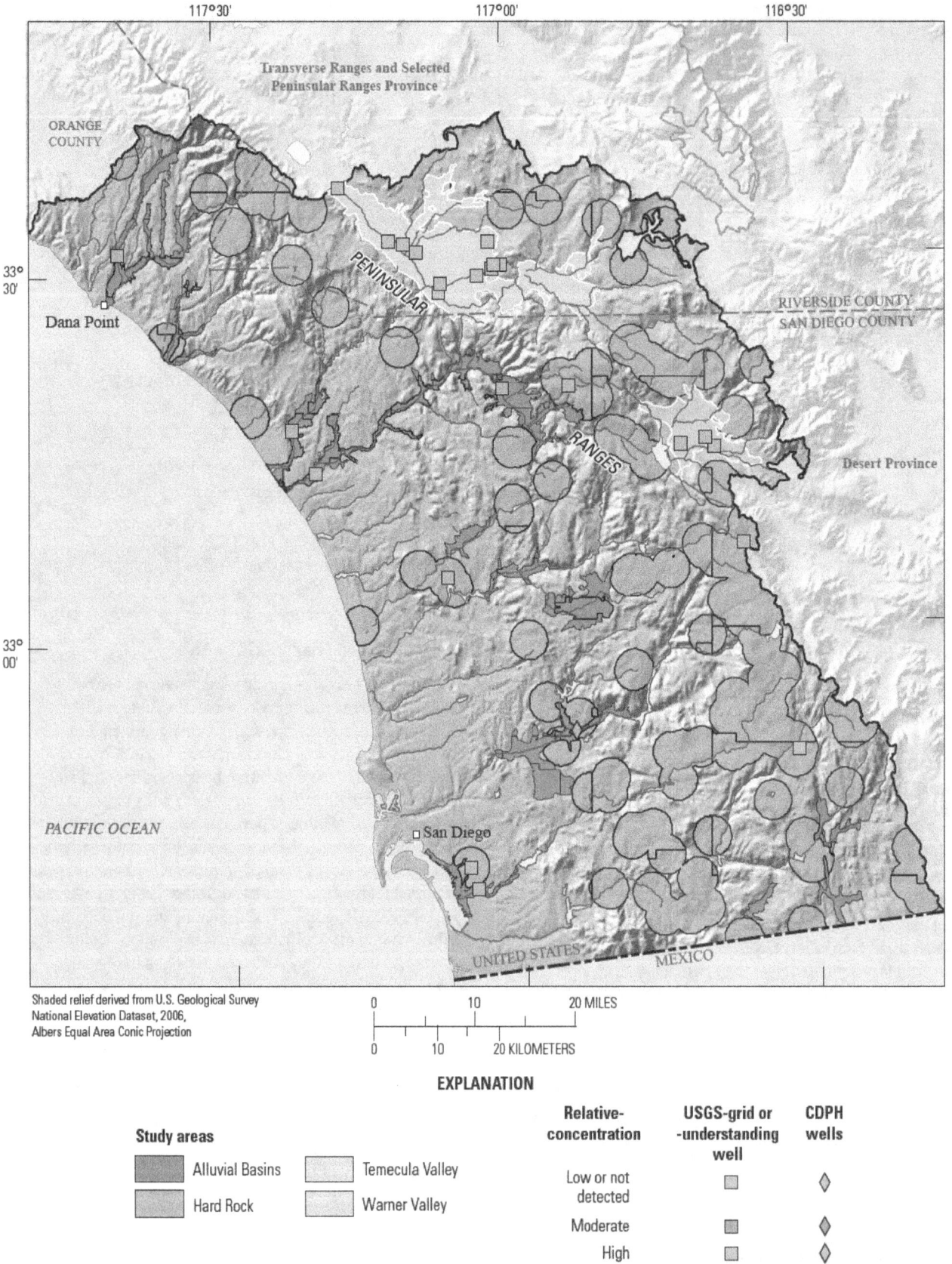

Shaded relief derived from U.S. Geological Survey
National Elevation Dataset, 2006,
Albers Equal Area Conic Projection

EXPLANATION

Study areas		Relative-concentration	USGS-grid or -understanding well	CDPH wells
Alluvial Basins	Temecula Valley	Low or not detected		
Hard Rock	Warner Valley	Moderate		
		High		

Figure 19. Values of radon-222 in USGS-grid and -understanding wells representative of the primary aquifers and the most recent analysis July 30, 2001–July 29, 2004, for CDPH wells, San Diego Groundwater Ambient Monitoring and Assessment (GAMA) study unit, California, May–July 2004.

Organic and Special-Interest Constituents

Volatile organic compounds can be in paints, solvents, fuels, refrigerants, can be byproducts of water disinfection, and are characterized by their tendency to evaporate. In this report, VOCs are categorized as trihalomethanes, solvents, and gasoline components. Pesticides are used to control weeds, insects, or fungi in agricultural, urban, and suburban settings. In this report, pesticides are discussed only in terms of herbicides because those were only class of pesticides detected in grid wells in the San Diego study unit.

Maximum relative-concentration and detection frequency (in grid wells not area-weighted), at any concentration, were used as selection criteria for organic and special-interest constituents and are shown in figure 20. Seven organic and special-interest constituents met the selection criteria: chloroform, 1,2-Dichloropropane, methyl *tert*-butyl ether (MTBE), atrazine, simazine, prometon, and perchlorate (fig. 21). Overall, organic constituents were detected in 62 percent of the 47 grid wells (not area-weighted) in the San Diego study unit.

Of the 12 VOCs detected with a health-based benchmark, 10 were detected only at low relative-concentrations. One VOC, 1,2-Dichloropropane, was detected at moderate relative-concentration (fig. 21). MTBE was the only VOC detected at a concentration greater than a health-based benchmark. The trihalomethane (THM) chloroform was the only VOC detected in more than 10 percent of the grid wells (fig. 21). Overall, the detection frequency for VOCs in the 47 grid wells (not area-weighted) was 34 percent.

Of the 123 pesticides and pesticide degradates analyzed, 9 pesticides were detected in grid wells with a health-based benchmark (fig. 20). In addition to these pesticides, five pesticide degradation products (daughter compounds) were detected in grid wells. All concentrations of pesticides were less than health-based benchmarks. Three pesticides—simazine, prometon, and atrazine—were detected in greater than 10 percent of the grid wells sampled (fig. 21). Overall, the detection frequency (non-area-weighted) for pesticides with health based benchmarks was 45 percent, and for any pesticide or pesticide degradate the detection frequency was 53 percent.

Overall, organic constituents with health based benchmarks were detected at high relative-concentrations in 3.0 percent of the primary aquifers in the Alluvial Fill study areas (table 8). The Alluvial Basins was the only study area in which these constituents were detected at a high relative-concentration (tables B2A–C). No high relative-concentrations were detected in the Hard Rock study area (table B2D).

Trihalomethanes

The THMs chloroform and bromodichloromethane were detected in multiple wells in the Alluvial Fill study areas. All THMs were detected at low relative-concentrations. The detection frequencies for THMs in grid wells was highest in the Alluvial Basins study area (33.3 percent), and then in the Temecula Valley studies area (25 percent), but were not detected in grid wells in the Warner Valley or Hard Rock study areas (fig. 22). THMs were the most frequently detected class of VOCs in aquifers based on national assessments by the USGS National Water-Quality Assessment (NAWQA) Program (Zogorski and others, 2006).

Factors Affecting Trihalomethane Distribution

Potential sources of THMs include recharge from landscape irrigation with disinfected water, leakage from distribution or sewer systems, and various industrial and commercial sources (Ivahnenko and Barbash, 2004). On a national scale, the detection of THMs in groundwater is correlated to urban land use (Zogorski and others, 2006). In the San Diego study unit, the positive correlation of THMs to urban land use was significant (table 9). Figure 23A shows detection frequency and concentration of THMs in groundwater samples as a function of urban land use. Although THMs are most frequently detected in samples collected from wells located in areas where urban landuse is greater than 50 percent, the samples with the highest THM concentrations came from wells located in a non-urban area. These wells (SDTEM-10 and SDTEMFP-04) are tapping imported water used for engineered recharge, which may be the source of THMs in this area (fig. 23B).

Trihalomethane concentrations were significantly higher in wells with modern groundwater age classification than in wells with mixed and pre-modern groundwater age classifications (fig. 23B; table 9); significant differences between wells classified as mixed and pre-modern were not detected. Trihalomethanes were detected in 57 percent of the samples classified as modern; THM concentrations also were highest in samples collected from wells with a component of modern recharge. Although THM concentrations were not significantly correlated with depth to the top of the uppermost open interval, samples with the highest concentration were collected at wells with a depth of less than 100 ft (fig. 23B).

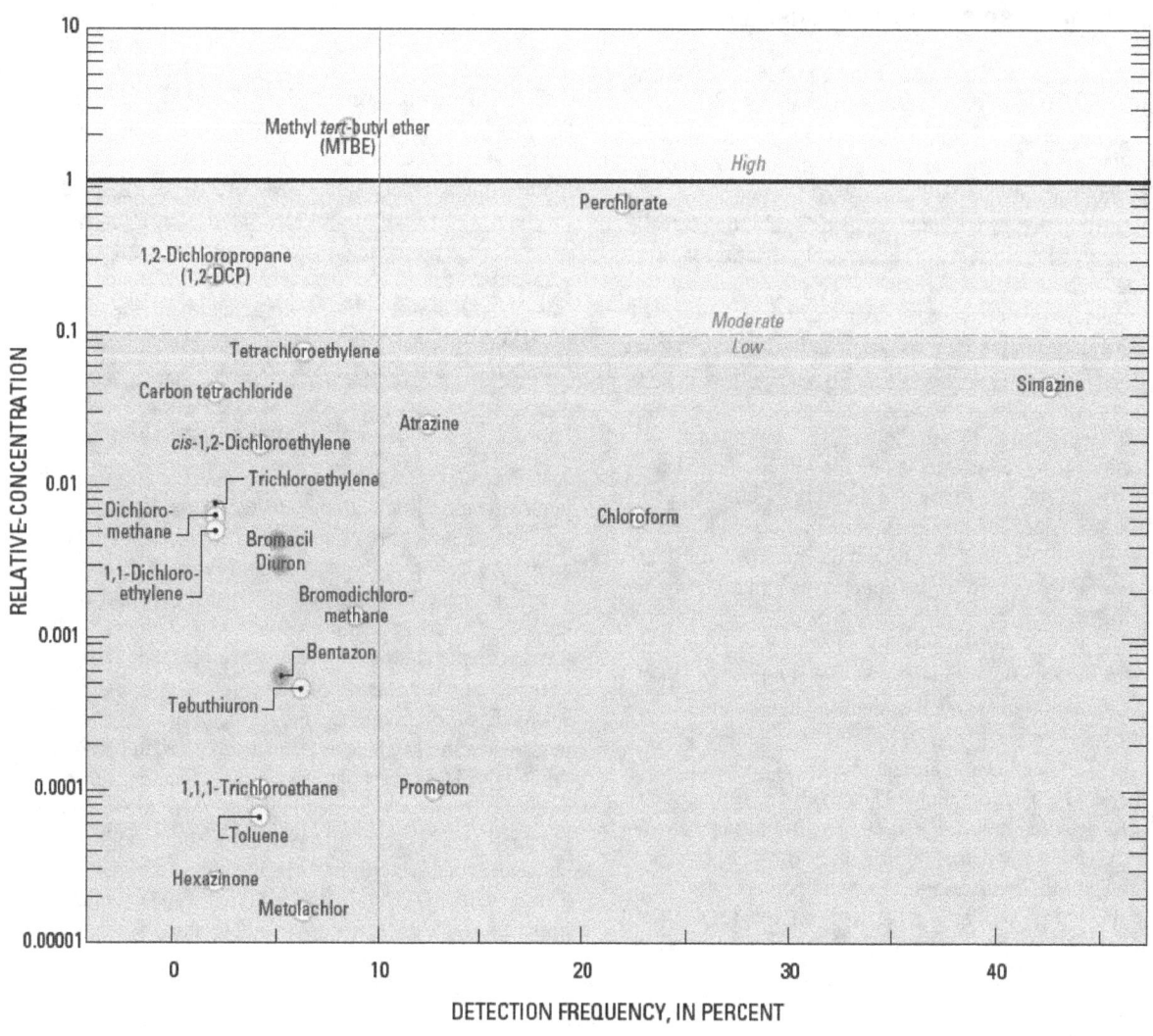

EXPLANATION

Hexazinone   **Constituents collected at all grid wells**—Name and center of symbol is location of data unless indicated
              by following location line:

Diuron   **Constituents collected at a subset of grid wells**—Name and center of symbol is location of data unless
              indicated by following location line:

Figure 20.   Detection frequency (non-area-weighted) and maximum relative-concentration for organic and special-
interest constituents detected in grid wells, San Diego Groundwater Ambient Monitoring and Assessment (GAMA)
study unit, California, May–July 2004.

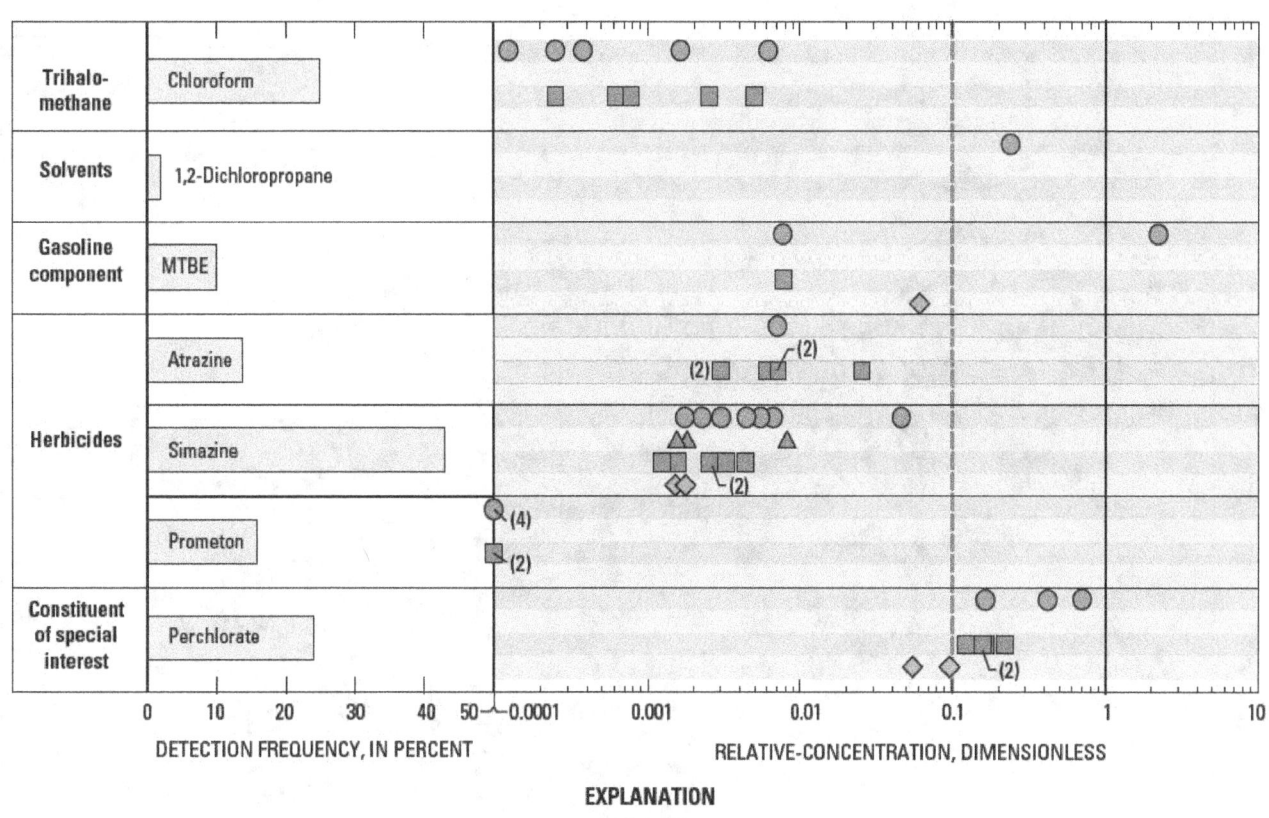

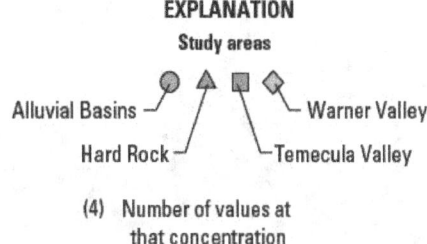

Figure 21. Detection frequency (non-area-weighted) and relative-concentrations in grid wells of selected organic and special-interest constituents, San Diego Groundwater Ambient Monitoring and Assessment (GAMA) study unit, California, May–July 2004.

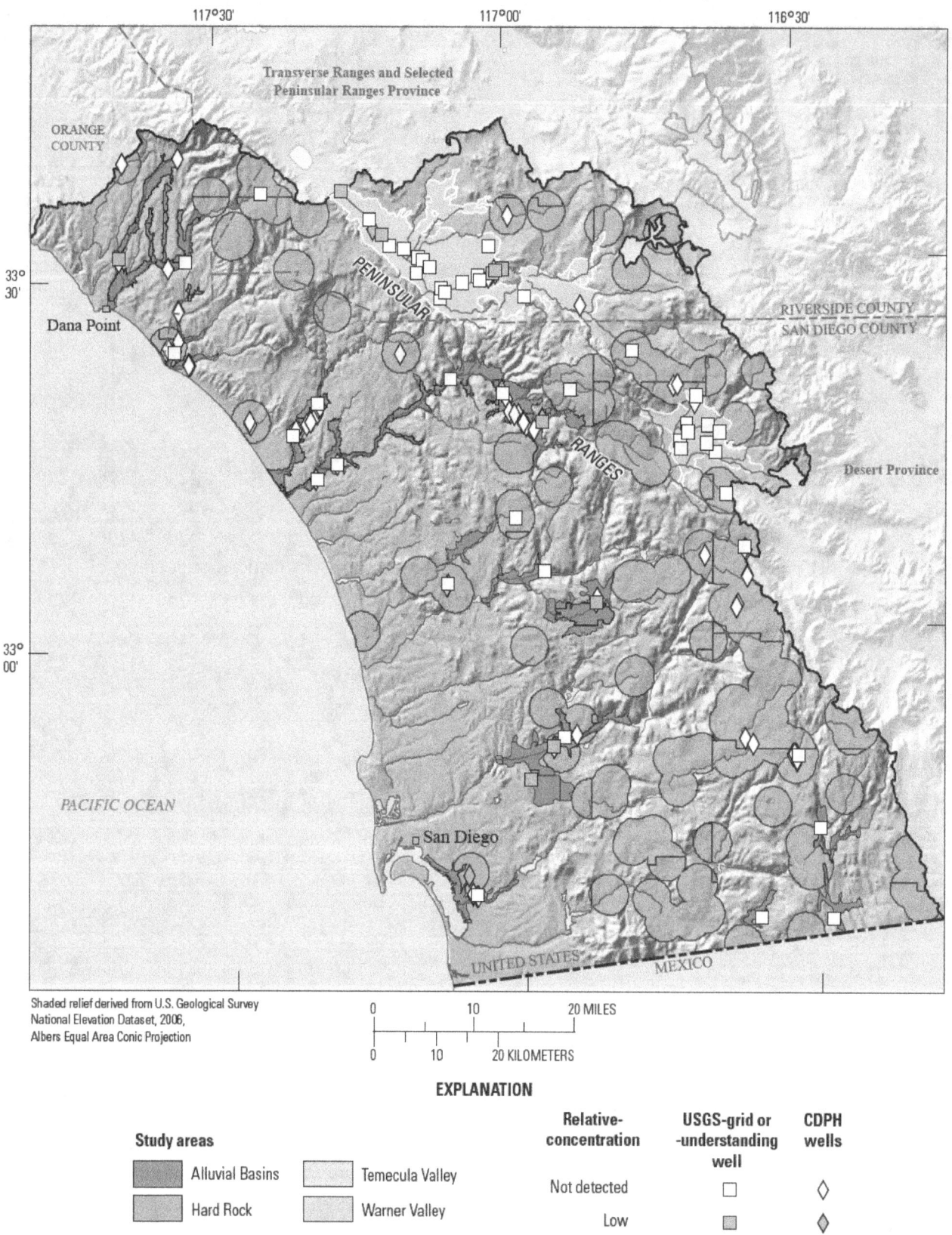

Figure 22.   Sum of trihalomethanes in USGS-grid and -understanding wells representative of the primary aquifers, and CDPH data from the prior 3-year period of study (July 30, 2001 to July 29, 2004), San Diego Groundwater Ambient Monitoring and Assessment (GAMA) study unit, California, May–July 2004.

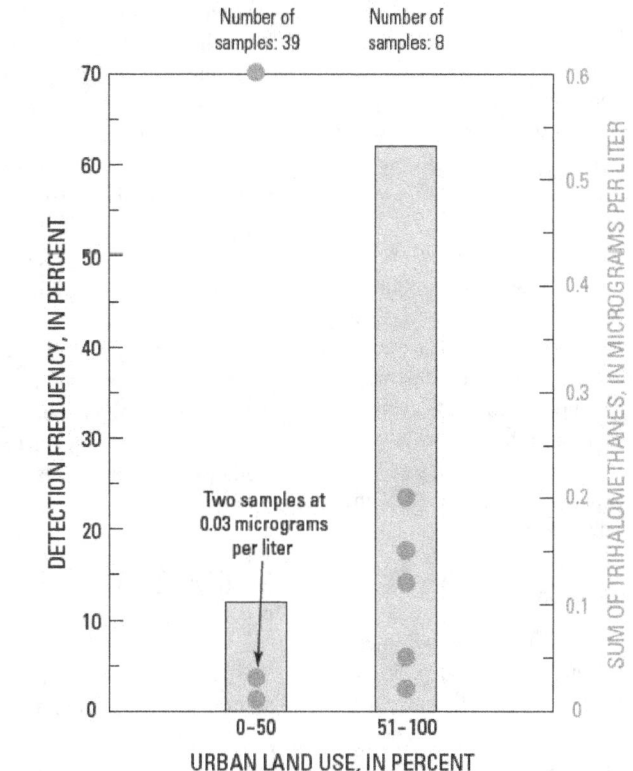

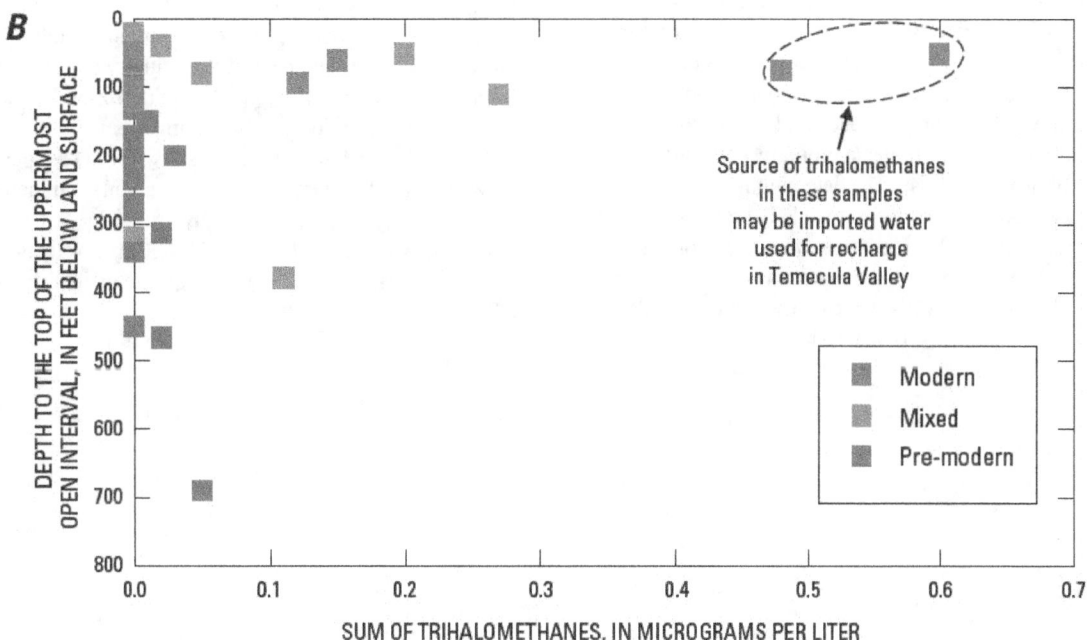

Figure 23. (*A*) Detection frequency and concentration of the trihalomethanes in relation to urban land use, and (*B*) concentration of trihalomethanes in relation to depth to the top of the uppermost open interval and groundwater age classification, San Diego Groundwater Ambient Monitoring and Assessment (GAMA) study unit, California, May–July 2004.

Solvents

The only solvent detected that met the selection criteria in the San Diego study unit was 1, 2-dichlororpane (fig. 20). In the CDPH database, 1, 2-dichloropropane was detected at high-relative concentrations in one well, but not during the current period (July30, 2001–July 29, 2004) (table 6). Two other solvents—tetrachloroethylene (PCE) and trichloroethylene (TCE)—were also detected at high-relative concentrations in the CDPH database, but not during the current period. Solvents were not detected at high relative-concentrations in the Alluvial Fill study areas, but were detected at moderate relative-concentrations in 3.0 percent of the primary aquifers (table 8). The detection frequency for solvents in grid wells was highest in the Temecula Valley study area (16.7 percent), and then the Alluvial Basins study area (13.3 percent), but was not detected in grid wells either in the Warner Valley or in the Hard Rock study areas (fig. 24).

Factors Affecting Solvent Distribution

Solvents are used for a variety of industrial, commercial, and domestic purposes (Zogorski and others, 2006). Solvents can be introduced into the subsurface through leaking storage tanks and disposal of waste streams from industrial and commercial processes. Nationally, solvent concentrations have been correlated with urban land-use (Zogorski and others, 2006; Moran and others, 2007). Like THM concentrations, a positive correlation of the sum of solvent concentrations to urban land-use was significant in the San Diego study unit (table 9). The sum of solvents was calculated from the summation of concentrations of all four solvents detected—PCE, TCE, 1,2-dichloropropane, and carbon tetrachloride. Figure 25 shows detection frequency and sum of the detected solvent concentrations in groundwater samples as a function of urban land-use. The detection frequency for solvents in samples collected from wells in areas with urban land-use greater than 50 percent was 38 percent compared to the detection frequency of just 5 percent for solvents in samples collected from wells in areas where the urban land use was 50 percent or less. In addition, the sum of solvent concentrations was higher in samples collected in urbanized areas than in non-urbanized areas.

The sum of solvent concentrations was not significantly different between groundwater age classes, or between wells with varying depth to the top of the uppermost open interval (table 9). Solvents were more frequently detected in pre-modern water (16 percent) than either in modern (11 percent) or in mixed (12 percent) waters. It is expected that solvents would be more prevalent in younger rather than older groundwater because these compounds most likely were used more in the last 60 years or so. However, because some solvents were used before 1950, it is plausible that these compounds could be present in pre-modern water. Additionally, solvents in pre-modern water could indicate short-circuit mechanisms resulting from well construction, well operation processes, or other non-advective transport processes.

Methyl *Tert*-Butyl Ether (MTBE) and Gasoline Components

MTBE was the only gasoline component detected that met the selection criteria in the San Diego study unit (fig. 26). In the CDPH database, MTBE was detected at high-relative concentrations in two wells; one detection was during the current period (July30, 2001–July 29, 2004) (table 4). The only other gasoline component detected was benzene; a single detection, high-relative concentration of this compound was recorded in the CDPH database during the current period. Gasoline components were detected at high relative-concentrations in 3.0 percent of the primary aquifers in the Alluvial Fill study areas (table 8). The detection frequency at any concentration for gasoline components in grid wells was highest in the Alluvial Basins study area (18.8 percent), followed by the Warner Valley (11.1 percent) and then by the Temecula Valley (8.3 percent) study areas (fig. 26). Gasoline components were not detected in USGS-grid wells in the Hard Rock study area, but were detected in five wells listed in the CDPH.

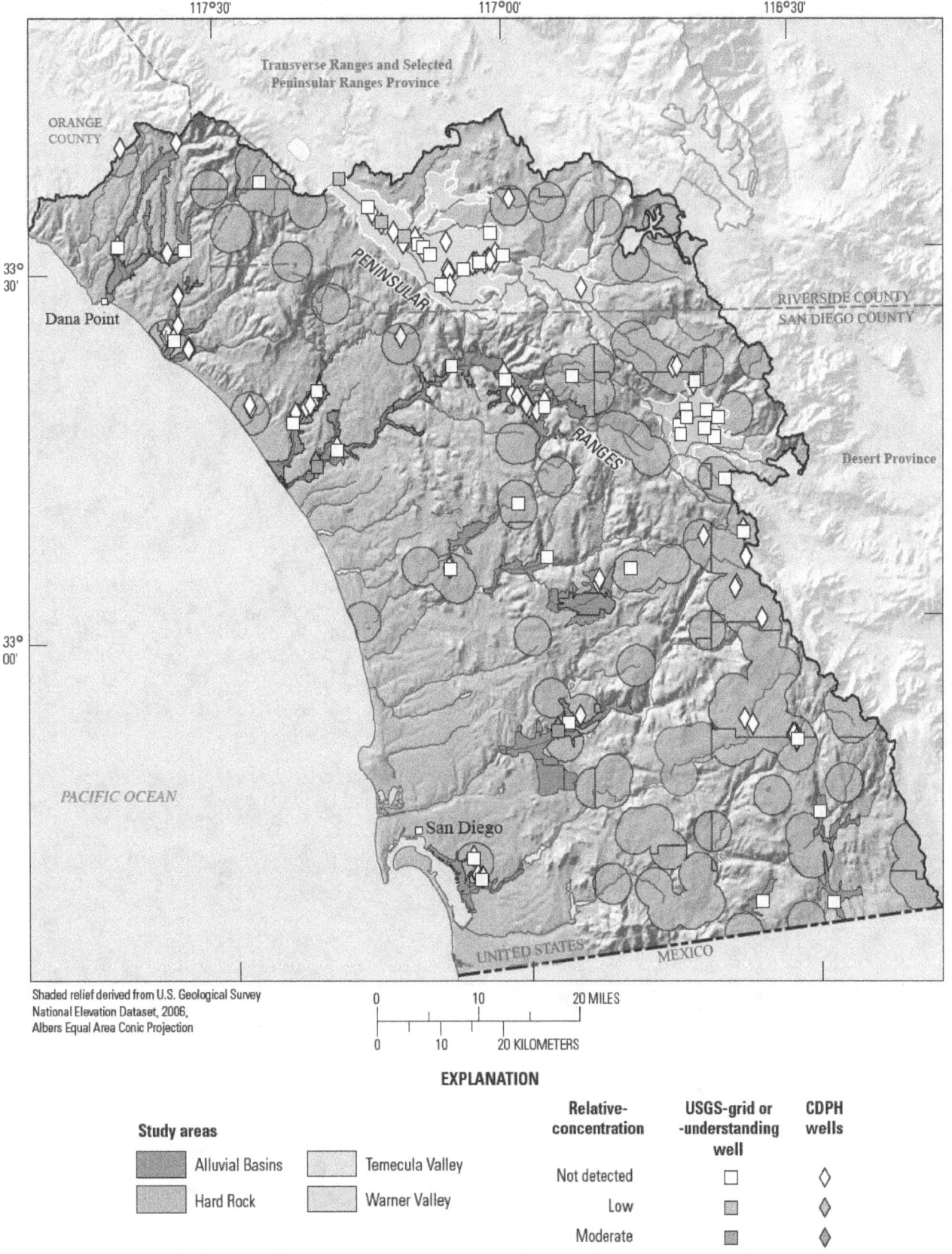

117°30'   117°00'   116°30'

Transverse Ranges and Selected
Peninsular Ranges Province

ORANGE
COUNTY

33°
30'

Dana Point

RIVERSIDE COUNTY
SAN DIEGO COUNTY

PENINSULAR

RANGES

Desert Province

33°
00'

PACIFIC OCEAN

San Diego

UNITED STATES   MEXICO

Shaded relief derived from U.S. Geological Survey
National Elevation Dataset, 2006,
Albers Equal Area Conic Projection

0   10   20 MILES

0   10   20 KILOMETERS

EXPLANATION

Study areas		Relative-concentration	USGS-grid or -understanding well	CDPH wells

Study areas

Alluvial Basins   Temecula Valley

Hard Rock   Warner Valley

Relative-concentration	USGS-grid or -understanding well	CDPH wells
Not detected	□	◇
Low	▨	◈
Moderate	▦	◆

Figure 24.   Sum of the solvents 1,2-dichloropropane, tetrachloroethylene (PCE), trichloroethylene (TCE), and carbon tetrachloride in USGS-grid wells representative of the primary aquifers, and CDPH data from the current period (July 30, 2001–July 29, 2004), San Diego Groundwater Ambient Monitoring and Assessment (GAMA) study unit, California, May–July 2004.

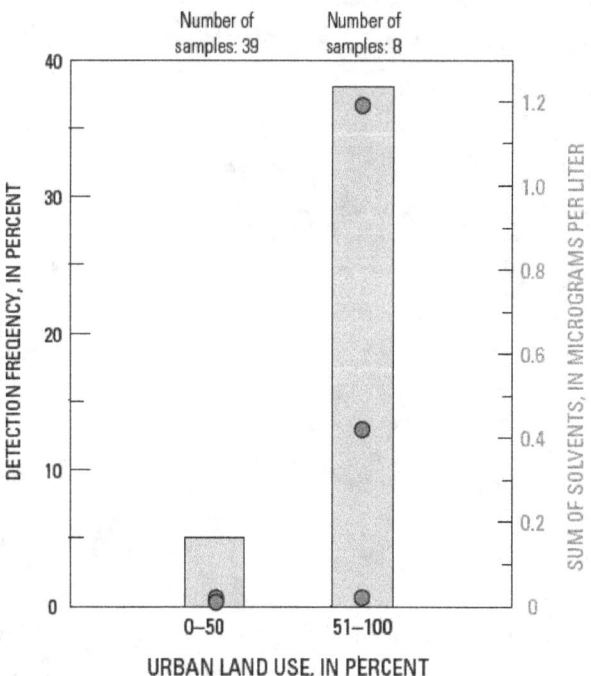

Figure 25. Detection frequency and sum of solvents concentration in relation to urban land use, San Diego Groundwater Ambient Monitoring and Assessment (GAMA) study unit, California, May–July 2004.

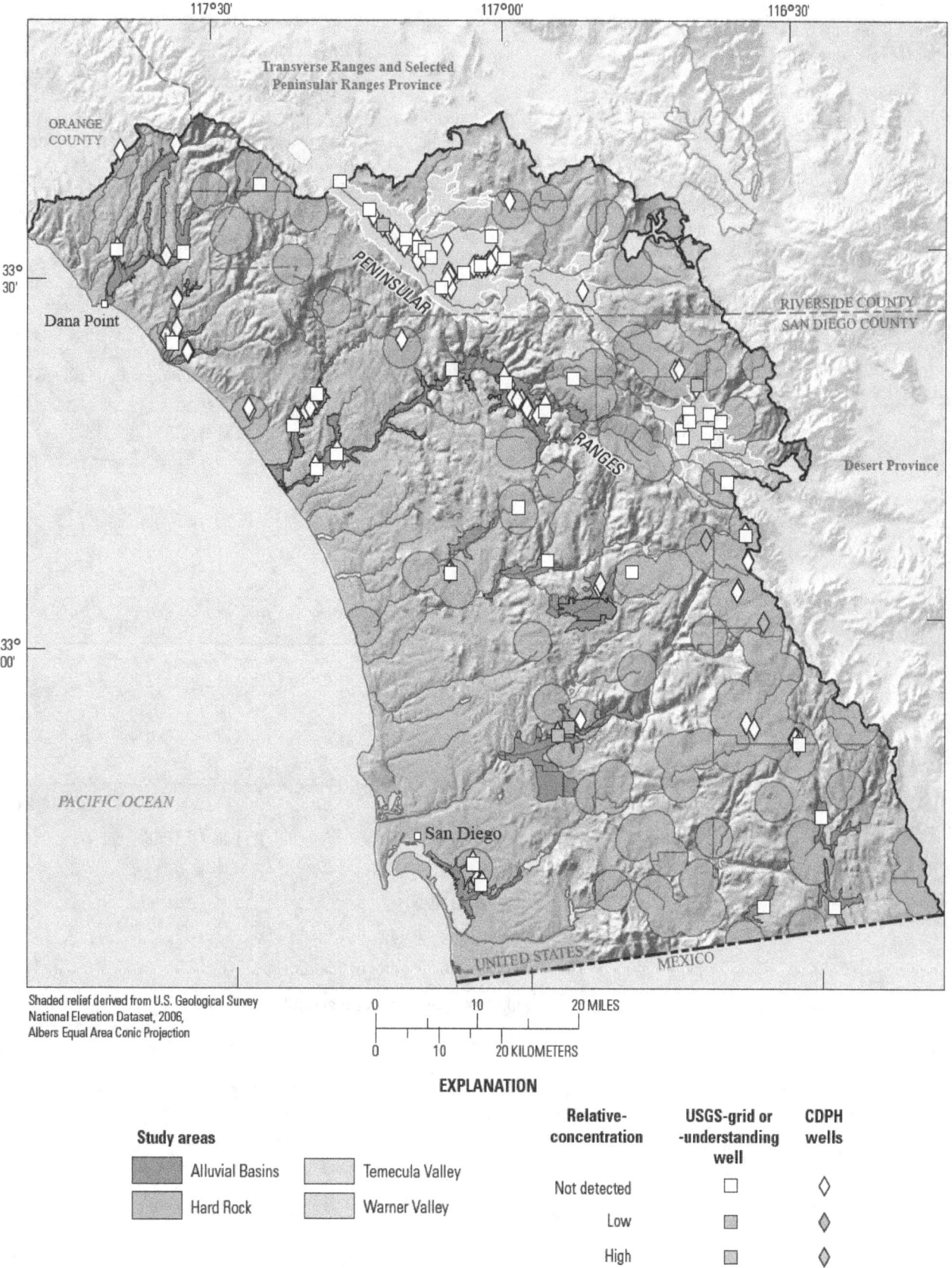

117°30'   117°00'   116°30'

Transverse Ranges and Selected
Peninsular Ranges Province

ORANGE
COUNTY

33°
30'

Dana Point

PENINSULAR

RIVERSIDE COUNTY
SAN DIEGO COUNTY

Desert Province

RANGES

33°
00'

PACIFIC OCEAN

San Diego

UNITED STATES   MEXICO

Shaded relief derived from U.S. Geological Survey
National Elevation Dataset, 2006,
Albers Equal Area Conic Projection

0   10   20 MILES

0   10   20 KILOMETERS

EXPLANATION

Study areas		Relative-concentration	USGS-grid or -understanding well	CDPH wells

Study areas

Alluvial Basins   Temecula Valley

Hard Rock   Warner Valley

Relative-concentration	USGS-grid or -understanding well	CDPH wells
Not detected	□	◇
Low	▨	◈
High	▦	◆

Figure 26. Sum of the gasoline components methyl *tert*-butyl ether (MTBE) and benzene in USGS-grid wells and from CDPH data for the prior 3-year period of study (July 30, 2001–July 29, 2004), San Diego Groundwater Ambient Monitoring and Assessment (GAMA) study unit, California.

Factors Affecting the Distribution of Methyl *Tert*-Butyl Ether (MTBE) and other Gasoline Components

Gasoline components, in particular MTBE, can be introduced to the sub-surface environment through several pathways. These components may be released into groundwater from point sources, such as leaking underground fuel tanks (LUFT), (Zogorski and others, 2006; Moran and others, 2007), or non-point sources, such as urban precipitation and storm water runoff (Pankow and others, 1997; Moran and others, 1999). Previous water-quality studies in California have indicated both point and non-point sources as the source of MTBE in groundwater (Happel and others, 1998; Belitz and others, 2003; Moran and others, 2004).

The sum of gasoline components was not significantly correlated to any of the explanatory factors listed in table 9. An additional analysis was done by comparing MTBE concentrations in groundwater samples to the distance from the nearest LUFT. Data for the LUFTs were obtained from the California State Water Resource Control Board's Geotracker Database (State Water Resources Control Board, 2011). The negative correlation of MTBE concentrations to distance from the nearest LUFT was significant (ρ = -0.54) (fig. 27). Of the 13 wells sampled within 500 m of a LUFT, MTBE was detected at 62 percent; MTBE was not detected in any sample collected greater than 500 m from a well. These results suggest that LUFTs are the primary source of MTBE detected in groundwater in the San Diego study unit. It must be noted however, that MTBEs were not detected in five wells sampled within 500 m of a LUFT. Non-detections in these wells may be a result of several factors, such as MTBE not being a component of the liquid "contained" within the LUFT, severity of the leak, well pumping intensity, dilution with unaffected water, and (or) rates of biodegradation.

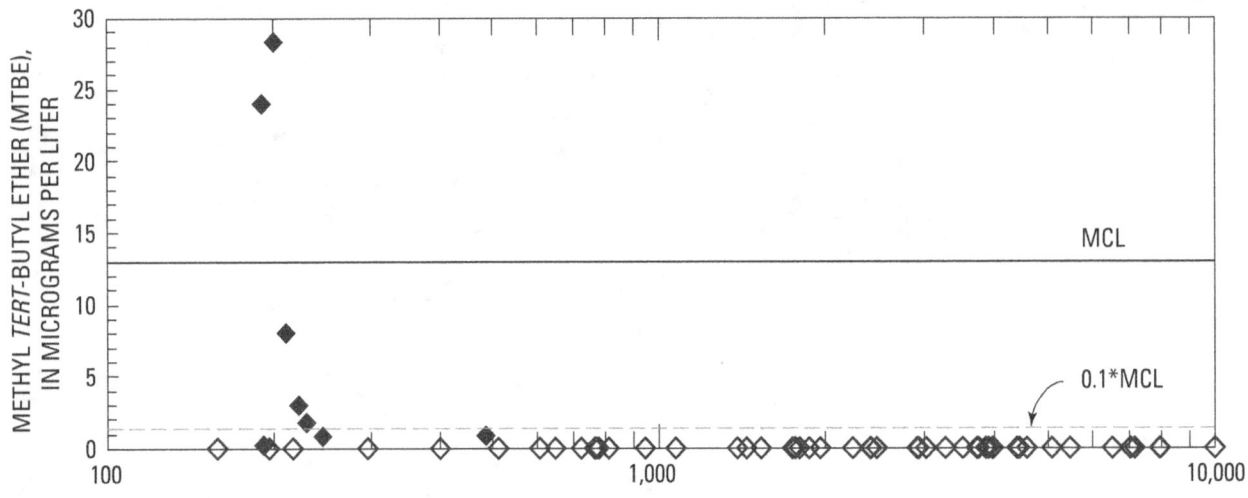

DISTANCE TO NEAREST LEAKING UNDERGROUND FUEL TANKS (LUFT), IN METERS

EXPLANATION

◆   Methyl *tert*-butyl ether detected

◇   Methyl *tert*-butyl ether not detected

—   Maximum Contaminant Level (MCL)

– – –   0.1*MCL

Figure 27.   Relation of methyl *tert*-butyl ether (MTBE) concentrations to distance from nearest leaking underground fuel tank (LUFT), San Diego Groundwater Ambient Monitoring and Assessment (GAMA) study unit, California, May–July 2004.

Pesticides

The only pesticides sampled for in the San Diego study unit that met the selection criteria were simazine, atrazine, prometon. Results from a study of major aquifers across the United States showed that these three compounds were frequently detected in groundwater (Gilliom and others, 2006). A groundwater study conducted in California showed that simazine was the most frequently detected triazine herbicide in groundwater (Troiano and others, 2001). Figure 28 shows the distribution of the sum of herbicide concentrations detected in the San Diego study unit. All detections of herbicides were observed at low relative-concentrations in the Alluvial Fill study areas (table 8). Herbicides most frequently were detected in grid wells in the Temecula Valley study area (66.7 percent), followed by grid wells in the Alluvial Basins (62.5 percent), in the Hard Rock (40 percent), and in the Warner Valley (22.2 percent) study areas.

Factors Affecting Pesticide Distribution

Simazine and prometon frequently are used for nonagricultural applications including weed control on bare ground, around buildings, along roadsides, and in other right-of-ways. Simazine also is used on a variety of crops including citrus and vineyards, whereas prometon has no registered agricultural uses (Gilliom and others, 2006). On the other hand, atrazine mostly is used for agricultural purposes in the control of weeds in row crops; some use is reported for the control of weeds in right of ways. The sum of herbicide concentrations was not significantly correlated to any land-use type in the San Diego study unit (table 9).

Concentrations of herbicides were highest in shallow wells with modern and mixed groundwater ages (fig. 29); the correlation between depth to the top of the uppermost open interval (at the 90 percent confidence level) and ground-water age classification are significant (table 9). Herbicides primarily were detected in wells with depths to the top of the open interval less than 200 ft. The reason that the correlation of herbicides to shallow well depths is stronger than are that of THMs, solvents, and gasoline components is likely because the soil organic carbon/water partition coefficient (K_{oc}) values for herbicides are higher than those values for VOCs. Compounds with relatively high K_{oc} are hydrophobic and are more likely to accumulate in soil and sediment than compounds with a low K_{oc} which are more likely to be dissolved in water. Therefore, herbicides as a result of a high K_{oc} tend to be less readily transported through soil and into groundwater.

Some herbicides were detected in relatively deep wells that are tapping pre-modern water. Detections of herbicides in relatively deep wells with pre-modern groundwater ages potentially could be influenced by short-circuiting mechanisms that allow small quantities of modern water with dissolved herbicides to mix with herbicide-free pre-modern water. Results from a USGS study in the San Joaquin Valley suggests that inter-borehole flow may cause mixing of shallow and deep groundwater during non-pumping conditions (Jurgens and others, 2008).

Herbicide concentrations also were significantly correlated with redox conditions (fig. 29B; table 9). Concentrations were higher in samples collected from wells tapping oxic waters with relatively low pH. Although geochemical parameters such as oxygen content and pH of groundwater may possibly affect the distribution of herbicides, this may not be the cause for the distribution of herbicides reported in this study. The correlation observed for oxic water and pH may result from shallow wells that tend to tap groundwater with a lower pH (table 7) and higher oxygen content than deeper wells

Perchlorate and Special-Interest Constituents

Constituents of special interest analyzed for in the San Diego study unit were NDMA, 1, 4-dioxane, and perchlorate. These constituents were selected because they recently were detected in, or are considered to have the potential to reach, drinking-water supplies (California Department of Public Health, 2008 b,c,d). NDMA and 1-4-dioxane were not detected in any wells (Wright and others, 2005). However, perchlorate was detected in eight grid wells and three CDPH wells (fig. 30). In the CDPH database perchlorate was detected as high in one well during the current period (July 30, 2001–July 29, 2004) (table 4). Perchlorate was detected at high relative-concentrations in 0.2 percent (calculated using spatially weighted approach) of the primary aquifers in the Alluvial Fill study areas (table 8). Perchlorate most frequently was detected in grid wells in the Temecula Valley study area (33.3 percent), followed by grid wells in the Warner Valley (22.2 percent) and in the Alluvial Basins (18.8) study areas; perchlorate was not detected in grid wells in the Hard Rock study area.

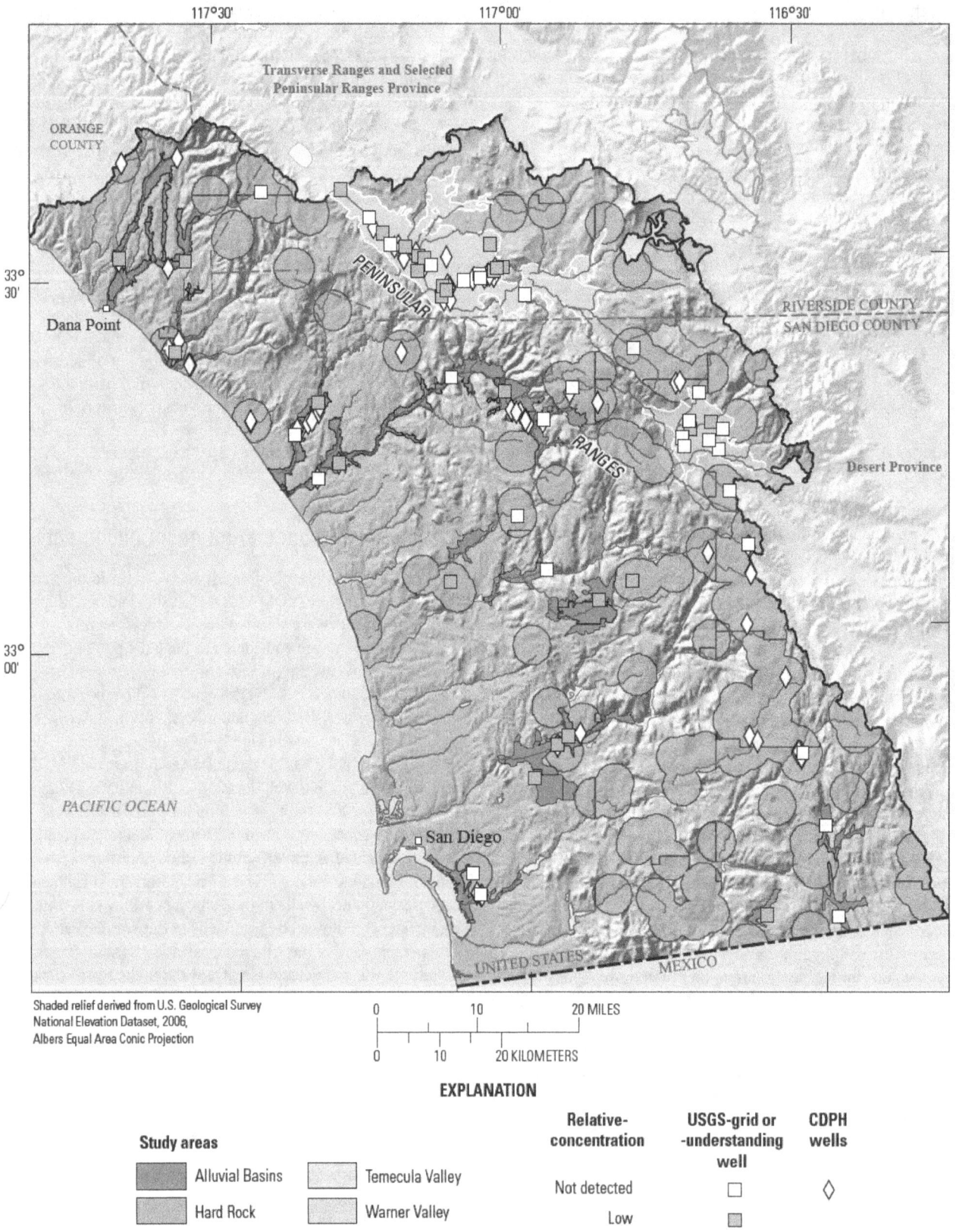

Figure 28.   Sum of herbicides in USGS-grid wells and from CDPH data for the prior 3-year period of study (July 30, 2001–July 29, 2004), San Diego Groundwater Ambient Monitoring and Assessment (GAMA) study unit, California, May–July 2004.

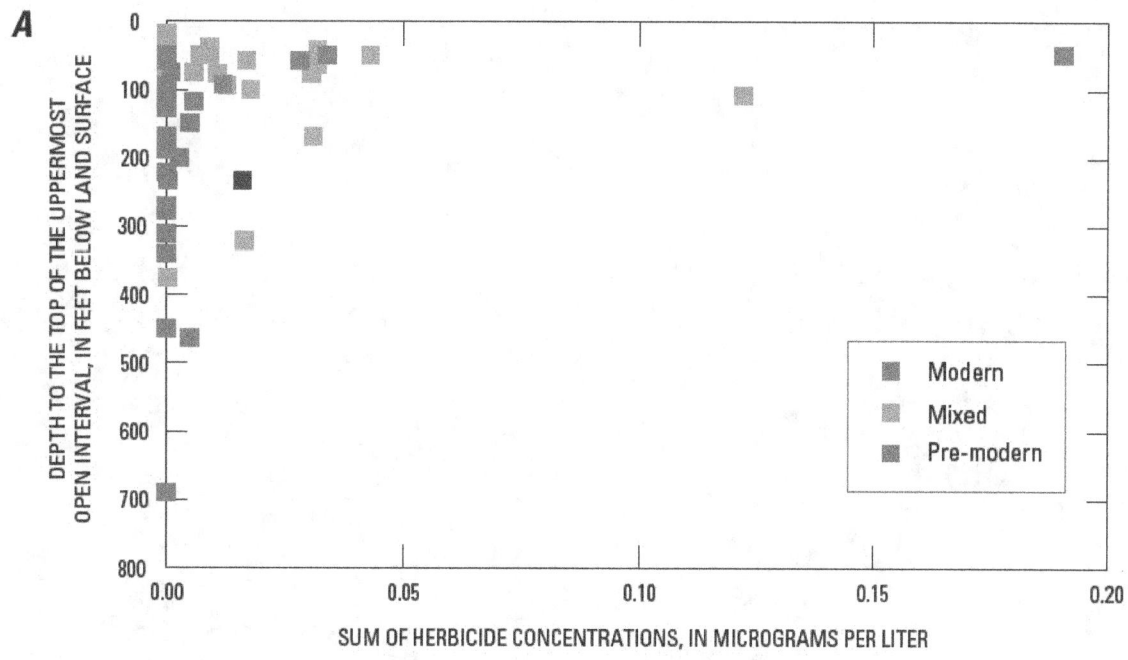

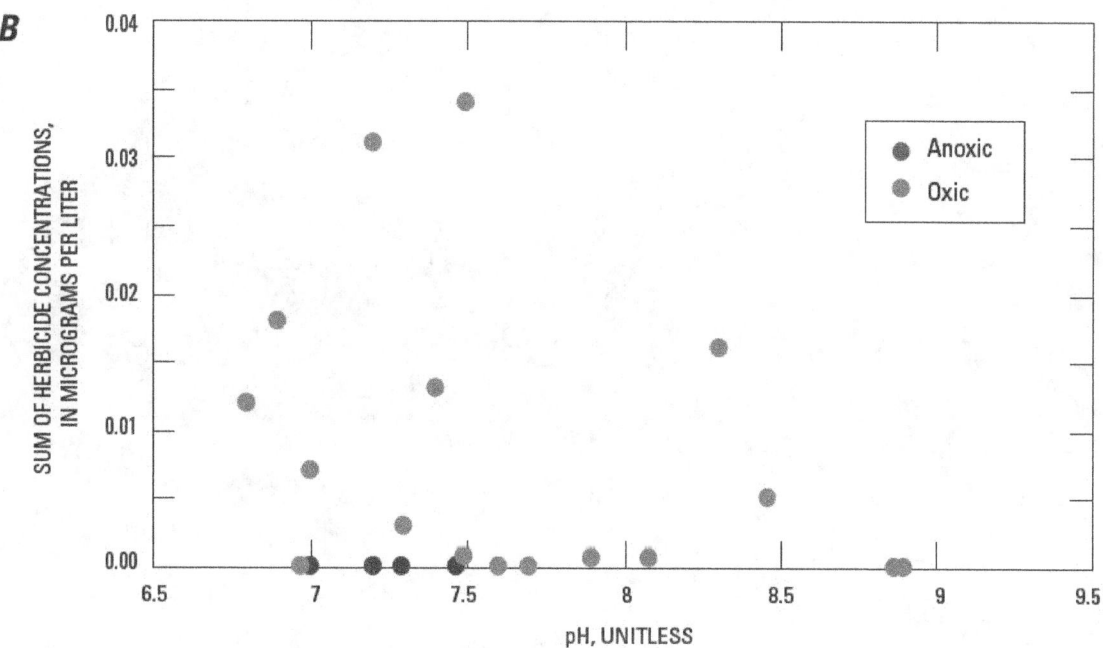

Figure 29. Sum of herbicide concentrations in relation to (*A*) depth to the top of the uppermost open interval and groundwater age classification, and in relation to (*B*) redox condition of groundwater, San Diego Groundwater Ambient Monitoring and Assessment (GAMA) study unit, California, May–July 2004.

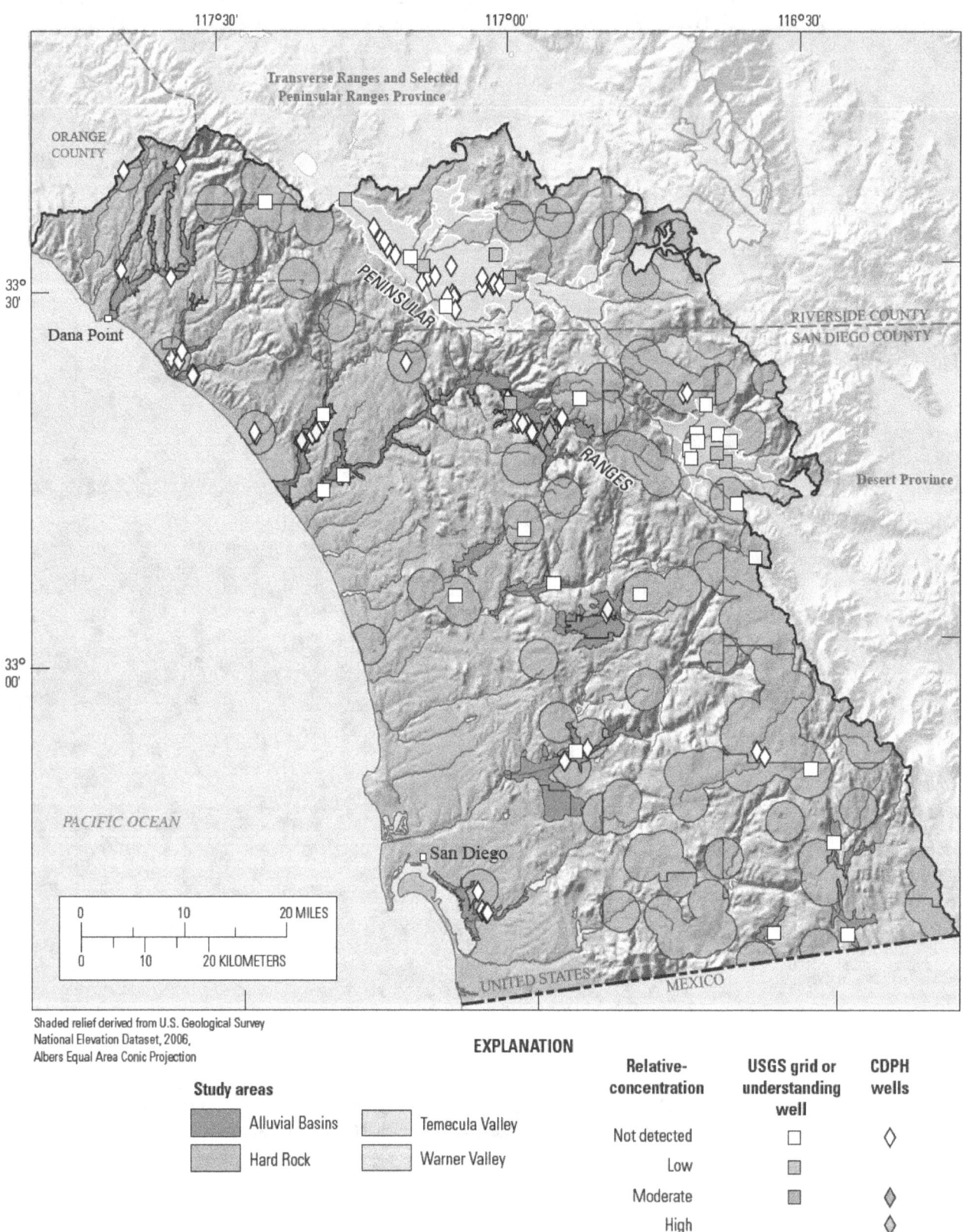

Shaded relief derived from U.S. Geological Survey
National Elevation Dataset, 2006,
Albers Equal Area Conic Projection

EXPLANATION

Study areas

▨	Alluvial Basins	▨	Temecula Valley
▨	Hard Rock	▨	Warner Valley

Relative-concentration	USGS grid or understanding well	CDPH wells
Not detected	□	◇
Low	▣	
Moderate	▣	◈
High		◈

Figure 30. Perchlorate in USGS-grid wells and from CDPH data for the prior 3-year period of study (July 30, 2001–July 29, 2004), San Diego Groundwater Ambient Monitoring and Assessment (GAMA) study unit, California, May–July 2004.

Factors Affecting Perchlorate Distribution

Potential anthropogenic sources of perchlorate include nitrate fertilizers (Dasgupta and others, 2006) and the production and use of explosives, road flares, and automobile air-bag systems (Parker and others, 2008). In addition, Colorado River water, which is imported into the San Diego study unit for public supply and agricultural irrigation, is known to contain perchlorate (California Department of Public Health, 2008b). Perchlorate also has been detected in groundwater in some highly arid desert environments as a result of natural atmospheric and soil processes (Dasgupta and others, 2005; Plummer and others, 2006); However, it is unclear whether these processes are occurring, or have occurred, in the San Diego study unit.

Perchlorate concentrations were only significantly correlated (negatively) to natural land-use in this study (table 9). Perchlorate concentrations, however, were positively correlated to agricultural areas at the 90 percent confidence level indicating that the use of nitrate fertilizers, and (or) Colorado River water that is used for irrigation, are contributing sources of perchlorate in groundwater of the San Diego study unit. Additionally, a detection of perchlorate (which is an order of magnitude below the MCL-CA) may be the result of a sample collected from a well (SDTEM-10) that is located down-gradient from an engineered recharge facility that uses Colorado River water.

Summary

The Groundwater Ambient Monitoring and Assessment (GAMA) Program was created by the California State Water Resources Control Board (State Water Board) to provide a comprehensive groundwater-quality baseline for the State of California. The program is a comprehensive assessment of statewide groundwater quality designed to improve ambient groundwater-quality monitoring and to increase the availability of information about groundwater quality to the public. The GAMA program includes the Priority Basin Project, conducted by the U.S. Geological Survey (USGS) in collaboration with the State Water Board and Lawrence Livermore National Laboratory (LLNL). This report is one of a series of reports presenting the *status* and *understanding* of current groundwater-quality conditions in study units of the GAMA Priority Basin Project.

The approximately 3,900-square mile (mi^2) San Diego study unit lies in the southwestern-most corner of California and is composed of four study areas—Temecula Valley (140 mi^2), Warner Valley (37 mi^2), Alluvial Basins (166 mi^2), and the Hard Rock (850 mi^2). The purposes of this report are (1) to describe briefly the hydrogeologic setting of the San Diego study unit, (2) to assess the current status of untreated groundwater quality in the primary aquifers in the San Diego study unit, and (3) to assess the relations between water quality and selected potential explanatory factors.

The GAMA San Diego study was designed to provide a statistically robust assessment of untreated groundwater quality within the primary aquifer systems. The primary aquifers were defined by the depth interval of the wells listed in the California Department of Public Health (CDPH) database for the San Diego study unit. Forty-seven grid wells were selected randomly within spatially distributed grid cells across the San Diego study unit. Grid wells were selected for sampling in each study area: 12 in the Temecula Valley, 9 in the Warner Valley, 16 in the Alluvial Basins, and 10 in the Hard Rock. In addition, 23 CDPH wells with data from the prior 3-year period (July 30, 2001–July 29, 2004) were used to complement USGS data. Grid-based and spatially weighted approaches were used to assess aquifer-scale proportions of constituents at high, moderate, and low relative-concentrations in the primary aquifers in order to characterize the quality of untreated groundwater in the study unit. Area weighting was used to account for the disparate size of the study areas relative to each other.

Given the large number of analytes, an objective algorithm was used to select those constituents of greatest importance to water quality in the primary aquifers for discussion in the report. To provide context, concentrations of constituents measured in the untreated groundwater were compared with regulatory and non-regulatory human-health and aesthetic benchmarks. Relative-concentrations (sample concentration divided by benchmark concentration) were used as the primary metric for evaluating groundwater quality. Constituents were classified into those whose maximum relative-concentrations were high, moderate, or low. Inorganic constituents with high relative-concentrations in greater than or equal to 2 percent of the grid wells (based on non area-weighted detections for all study areas in the San Diego study unit) were tested for relations to a set of potential explanatory factors that included land use, classified groundwater age, and geochemical-condition indicators. For organic and special-interest constituents to be tested for relations with potential explanatory variables detection frequencies had to be greater than 10 percent in grid wells (non area-weighted) at any concentration, or be detected at least once in a grid well at a moderate or high relative-concentration.

Inorganic constituents with human-health benchmarks were high in 17.6 percent of the primary aquifers in the Alluvial Fill study areas. Aquifer-scale high proportions of inorganic constituents with human-health based benchmarks indicated high relative-concentrations of trace elements, nutrients, and radioactive constituents. Trace elements with high concentrations in greater than or equal to 2 percent of the primary aquifers in the San Diego study unit were vanadium (V), arsenic (As), and boron (B). Inorganic constituents with non-health based benchmarks at high relative-concentrations in greater than or equal to 2 percent in the primary aquifers in the San Diego study unit were manganese (Mn), iron (Fe), and total dissolved solids (TDS).

The relation between the concentrations of V, As, Mn, and Fe in groundwater and in urban or agricultural land-use was not significant. This result suggests that natural sources (dissolution of rocks) are the significant contributing factor of these constituents in groundwater. Concentrations of B and TDS had a significant positive correlation to land use. Boron was correlated with urban land use and TDS with agricultural land-use, thus indicating anthropogenic activities as significant sources of these constituents in groundwater in the San Diego study unit.

pH and redox conditions of groundwater were important factors that affect the concentrations of inorganic constituents in groundwater. Generally, concentrations of the trace elements V, As, and B were high in alkaline groundwater which suggests that these trace elements are being desorbed from, or are inhibited from adsorbing onto, mineral surfaces. Vanadium and As are redox-sensitive species, but only the correlation of V to redox conditions was significant. Vanadium concentrations were higher in oxic groundwater than in anoxic groundwater, indicating that in the San Diego study unit V is most mobile under oxic and alkaline conditions. Concentrations of V and As also were higher in deep wells rather than in shallow wells and in samples that were partially or entirely composed of pre-modern (> 50 yrs) groundwater. The correlations between concentrations and depth and groundwater age likely are because of the significant positive correlation of deep wells and pre-modern groundwaters to pH.

The negative correlation of Mn and Fe to pH and dissolved oxygen (DO) was statistically significant, a result of the differing mechanisms that enhance the mobility of Mn and Fe in groundwater. The negative correlation of TDS to pH also was significant; however, this relation likely is not the result of any geochemical chemical processes, but because the TDS concentrations were highest in shallow wells where the pH of groundwater is significantly lower than the pH of groundwater being tapped by deeper wells.

Organic and special-interest constituents were detected at high relative-concentrations in a smaller proportion of the primary aquifers in the Alluvial Fill study areas (3.0 percent) than were inorganic constituents (17.6 percent). Relative-concentrations of organic constituents were moderate in 3.0 percent of the primary aquifers in the Alluvial Fill study areas. The proportion of the primary aquifers with high relative-concentrations of organic constituents was due to a detection of the discontinued gasoline oxygenate MTBE. One organic and one special-interest constituent were each detected at moderate relative-concentrations in grid wells—1,2-dichloropropane and perchlorate.

The status assessment for organic constituents indicated that 12 of the 88 VOCs and 14 of the 123 pesticides and pesticide degradates analyzed in grid wells were detected. Of the 12 VOCs detected, 10 were detected at low relative-concentration, one at a moderate relative-concentration, and one at a high relative-concentration. Only one VOC, chloroform, was detected in greater than or equal to 10 percent of the grid wells. Of the nine herbicides with health-based benchmarks that were detected, all had low relative-concentrations. Detection frequencies for simazine, atrazine, and prometon were greater than or equal to 10 percent. Perchlorate was the only special interest constituent detected in groundwater samples; it was detected in 22 percent of samples.

The positive correlation of the sum of THMs and solvent concentrations to urban land-use was significant, indicating that in the San Diego study unit groundwater located underneath urbanized areas is more likely to have detections of these anthropogenic constituents. MTBE concentrations were negatively correlated to the distance from the nearest leaking underground fuel tank, indicating that these point sources are the most significant contributing factor for MTBE concentrations to groundwater in the San Diego study unit.

The positive correlation of concentrations of THMs and herbicides to modern groundwater was significant. The negative correlation of herbicides to pH and anoxic groundwater also was significant. Although pH and redox conditions may affect the distribution of herbicides in groundwater, the correlations observed in this study likely are because herbicides were detected more frequently and at high concentrations in shallow wells where groundwater conditions tend to be oxic with relatively low pH.

Acknowledgments

The authors thank the California State Water Board, Lawrence Livermore National Laboratory, the California Department of Public Health, and the California Department of Water Resources. We especially thank the well owners and water purveyors for their cooperation and generosity in allowing the USGS to collect samples from their wells. Funding for this work was provided by State of California bonds authorized by Proposition 50 and administered by the California State Water Board.

References

Aeschbach-Hertig, W., Peeters, F., Beyerle, U., and Kipfer, R., 1999, Interpretation of dissolved atmospheric noble gases in natural waters: Water Resources Research, v. 35, no. 9, p. 2779–2792.

Aeschbach-Hertig, W., Peeters, F., Beyerle, U., and Kipfer, R., 2000, Paleotemperature reconstruction from noble gases in ground water taking into account equilibration with entrapped air: Nature, v. 405, p. 1040–1044

Andrews, J.N., 1985, The isotopic composition of radiogenic helium and its use to study groundwater movement in confined aquifers: Chemical Geology, v. 49, p. 339–351.

Andrews, J.N., Lee, D.J., 1979, Inert gases in groundwater from the Bunter Sandstone of England as indicators of age and paleoclimatic trends: Journal of Hydrology, v. 41, p. 233–252.

Ayotte, J.D., Flanagan, S.M., and Morrow, W.S., 2007, Occurrence of uranium and radon-222 in glacial and bedrock aquifers in the northern United States: U.S. Geological Survey Scientific Investigations Report 2007-5037, 84 p. (Also available at http://pubs.usgs.gov/sir/2007/5037.)

Belitz, K., Dubrovsky, N.M., Burow, K.R., Jurgens, B., and Johnson, T., 2003, Framework for a groundwater quality monitoring and assessment program for California: U.S. Geological Survey Water-Resources Investigations Report 03-4166, 28 p

Belitz, K., Jurgens, B., Landon, M.K., Fram, M.S., and Johnson, T., 2010, Estimation of aquifer-scale proportion using equal-area grids: Assessment of regional-scale groundwater quality: Water Resources Research, v. 46, citation number W11550, doi:10.1029/2010WR009321, accessed March 16, 2011, at http://www.agu.org/pubs/crossref/2010/2010WR009321.shtml.

Brown, L.D., Cai, T.T., and DasGupta, A., 2001, Interval Estimation for a Binomial Proportion: Statistical Science, v. 16, no. 2, p.101–117.

California Department of Public Health, 2008a, California drinking water-related laws–Drinking water-related regulations, Title 22: California Department of Public Health website, accessed August 17, 2009, at http://www.cdph.ca.gov/certlic/drinkingwater/Pages/Lawbook.aspx.

California Department of Public Health, 2008b, Perchlorate in drinking water: California Department of Public Health website, accessed January 2, 2008, at http://www.cdph.ca.gov/certlic/drinkingwater/Pages/Perchlorate.aspx.

California Department of Public Health, 2008c, NDMA and other nitrosamines—Drinking water issues: California Department of Public Health website, accessed January 2, 2008, at http://www.cdph.ca.gov/certlic/drinkingwater/Pages/NDMA.aspx.

California Department of Public Health, 2008d, 1,2,3-Trichloropropane: California Department of Public Health website, accessed January 2, 2008, at http://www.cdph.ca.gov/certlic/drinkingwater/Pages/123TCP.aspx.

California Department of Water Resources, 1956, Santa Margarita River investigation: California Department of Water Resources Bulletin 57, 273 p.

California Department of Water Resources, 1971, Water wells in the San Luis Rey River Valley area, San Diego County, California: California Department of Water Resources Bulletin 91-18, 347 p.

California Department of Water Resources, 1991, San Diego Region ground water studies, Phase V—Ramona hydrologic subarea and Jamacha hydrologic subarea: Memorandum Report, 92 p.

California Department of Water Resources, 2003, California's groundwater: California Department of Water Resources Bulletin 118, 246 p. (Also available at http://www.water.ca.gov/groundwater/bulletin118/bulletin118update2003.cfm.)

California Department of Water Resources, 2004a, Temecula Valley Groundwater Basin: California Department of Water Resources Bulletin 118, accessed December, 2009, at http://www.water.ca.gov/pubs/groundwater/bulletin_118/basindescriptions/9-5.pdf.

California Department of Water Resources, 2004b, Warner Valley Groundwater Basin: California Department of Water Resources Bulletin 118, accessed December, 2009, at http://www.water.ca.gov/pubs/groundwater/bulletin_118/basindescriptions/9-8.pdf.

California Department of Water Resources, 2004c, Potrero Valley Groundwater Basin: California Department of Water Resources Bulletin 118, accessed December, 2009, at http://www.water.ca.gov/pubs/groundwater/bulletin_118/basindescriptions/9-29.pdf.

California Department of Water Resources, 2004d, San Luis Rey Valley Groundwater Basin: California Department of Water Resources Bulletin 118, accessed December 2009, at http://www.water.ca.gov/pubs/groundwater/bulletin_118/basindescriptions/9-7.pdf.

California Regional Water Quality Control Board, San Diego Region, 1994, Water quality control plan for the San Diego Basin, v. 9, sec. 1, 3 p.

Clark, I.D., and Fritz, P., 1997, Environmental isotopes in hydrogeology: New York, Lewis Publishers, 328 p.

Cook, P.G., and Böhlke, J.K., 2000, Determining timescales for groundwater flow and solute transport, in Cook, P.G., and Herczeg, A., eds., Environmental tracers in subsurface hydrology: Boston, Kluwer Academic Publishers, p. 1–30.

Danskin, W.R., and Church, C.D., 2005, Determining age and vertical contribution of groundwater pumped from wells in a small coastal river basin—a case study in the Sweetwater River Valley, San Diego county, California: U.S. Geological Survey Open-File Report 2005-1032, 4 p.

Dasgupta, P.K., Dyke, J.V., Kirk, A.B., and Jackson, W., 2006, Perchlorate in the United States—Analysis of relative source contributions to the food chain: Environmental Science and Technology., v., 40, p. 6608–6614, doi:10/1021/es061321z, accessed March 9, 2011, at http://pubs.acs.org/doi/full/10.1021/es061321z.

Dasgupta, P.K., Martinelango, P.K., Jackson, W.A., Anderson,T.A., Tian, K., Tock, R.W., Rajagopalan, S., 2005, The origin of naturally occurring perchlorate—The role of atmospheric processes. Environmental Science and Technology, v. 39, p. 1569–1575, doi:10.1021/es048612x, accessed March 9, 2011, at http://pubs.acs.org/doi/abs/10.1021/es048612x.

Davis, S., and DeWiest, R.J., 1966, Hydrogeology: New York, John Wiley and Sons, 413 p.

Dotsika, E., Poutoukis D., Michelot, J.L., and Kloppmann, W., 2006, Stable isotope and chloride, boron study for tracing sources of boron contamination in groundwater—Boron contents in fresh and thermal water in different areas in Greece: Water, Air, and Soil Pollution, v. 174, p. 19–32.

Duce, R.A., and Hoffman, G.L., 1976. Atmospheric vanadium transport to the ocean: Atmospheric Environment, v. 10, no. 11, p. 989–996.

Fontes, J.C., and Garnier, J.M., 1979, Determination of the initial 14C activity of the total dissolved carbon: A review of the existing models and a new approach: Water Resources Research, v. 15, p. 399–413.

Gilliom, R.J., Barbash, J.E., Crawford, C.G., Hamilton, P.A., Martin, J.D., Nakagaki, N., Nowell, L.H., Scott, J.C., Stackelberg, P.E., Thelin, G.P., and Wolock, D.M., 2006, The quality of our Nation's waters—Pesticides in the Nation's streams and groundwater, 1992–2001: U.S. Geological Survey Circular 1291, 172 p.

Happel, A.M., Dooher, B., and Beckenbach, E.H., 1998, Methyl-tertiary butyl ether (MTBE) impacts to California groundwater: Livermore, Calif., Lawrence Livermore National Laboratory, UCRL-AR-130897, 68 p.

Helsel, D.R., and Hirsch, R.M., 2002, Statistical methods in Water Resources: U.S. Geological Survey Techniques of Water-Resources Investigations, book 4, chap. A3, 510 p. (Also available at http://water.usgs.gov/pubs/twri/twri4a3/.)

Hem, J.D., 1985, Study and interpretation of the chemical characteristics of natural water: U.S. Geological Survey Water-Supply Paper 2254, 213 p.

Hope, B.K., 1997, An assessment of the global impact of anthropogenic vanadium: Biogeochemistry, 37, no. 1, p. 1–13.

Isaaks, E.H., and Srivastava, R.M., 1989, Applied Geostatistics: New York, Oxford University Press, 511 p.

Ivahnenko, T., and Barbash, J.E., 2004, Chloroform in the hydrologic system—Sources, transport, fate, occurrence, and effects on human health and aquatic organisms: U.S. Geological Survey Scientific Investigations Report 2004-5137, 34 p. (Also available at http://pubs.usgs.gov/sir/2004/5137.)

Izbicki, J.A., 1985, Evaluation of the Mission, Santee, and Tijuana hydrologic subareas for reclaimed-water use, San Diego County, California: U.S. Geological Survey Water-Resources Investigations Report 85–4032, 99 p.

Johnson, T.D., and Belitz, K., 2009, Assigning land use to supply wells for the statistical characterization of regional groundwater quality—Correlating urban land use and VOC occurrence: Journal of Hydrology, v. 370, p. 100–108.

Jurgens, B.C., Burow, K.R., Dalgish, B.A., and Shelton, J.L., 2008, Hydrogeology, water chemistry, and factors affecting the transport of contaminants in the zone of contribution to a public-supply well in Modesto, eastern San Joaquin Valley, California: U.S. Geological Survey Scientific Investigations Report 2008–5156, 78 p. (Also available at http://pubs.usgs.gov/sir/2008/5156.)

Kalin, R.M., 2000, Radiocarbon dating of groundwater systems, in Cook, P.G., and Herczeg, A., eds., Environmental tracers in subsurface hydrology: Boston, Kluwer Academic Publishers, p. 111–144.

Kennedy, M.P., 1977, Recency and character of faulting along the Elsinore fault zone in southern Riverside County. California: Division of Mines and Geology Special Report 131.

Kulongoski, J., and Belitz, K., 2004, Groundwater ambient monitoring and assessment program: U.S. Geological Survey Fact Sheet 2004-3088.

Kulongoski, J.T., Hilton, D.R, Cresswell, R.G., Hostetler, S., and Jacobson, G., 2008, Helium-4 characteristics of groundwaters from Central Australia—Comparative chronology with chlorine-36 and carbon-14 dating techniques: Journal of Hydrology, v. 348, p. 176–194.

Landon, M.K., Belitz, K., Jurgens, B.C, Kulongoski, J.T., and Johnson, T.D., 2010, Status and Understanding of Groundwater Quality in the Central–Eastside San Joaquin Basin, 2006—California GAMA Priority Basin Project: U.S. Geological Survey Scientific Investigations Report 2009-5266, 97 p. (Also available at http://pubs.usgs.gov/sir/2009/5266/.)

Lucas, L.L., and Unterweger, M.P., 2000, Comprehensive review and critical evaluation of the half-life of tritium: Journal of Research of the National Institute of Standards and Technology, v. 105, no. 4, p. 541–549.

Manning, A.H., Solomon, D.K., and Thiros, S.A., 2005, ^{3}H/^{3}He age data in assessing the susceptibility of wells to contamination: Ground Water, v. 43, no. 3, p. 353–367.

McKelvey, V.E., Strobell, J.D., Jr., and Slaughter, A.L., 1986, The vanadiferous zone of the phosphoria formation in western Wyoming and southeastern Idaho (USA): U.S. Geological Survey Professional Paper 1465.

McMahon, P.B., and Chapelle, F.H., 2008, Redox processes and water quality of selected principal aquifer systems: Ground Water, v. 46, no. 2, p. 29–271.

Michel, R.L., 1989, Tritium deposition in the continental United States 1953–83: U.S. Geological Survey Water-Resources Investigations Report 89-4072, 46 p.

Michel, R., and Schroeder, R., 1994, Use of long-term tritium records from the Colorado River to determine timescales for hydrologic processes associated with irrigation in the Imperial Valley, California: Applied Geochemistry, v. 9, p. 387–401.

Moran, J.E., Hudson, G.B., Eaton, G.F., and Leif, R., 2004, California GAMA program—groundwater ambient monitoring and assessment results for the Sacramento valley and volcanic provinces of northern California: Lawrence Livermore National Laboratory, UCRL-TR-208179, 26 p.

Moran, M.J., Zogorski, J.S., and Squillace, P.J., 1999, MTBE in groundwater of the United States—Occurrence, potential sources, and long-range transport, in Water Resources Conference, Norfolk, Va., September 26–29, 1999, Proceedings: Denver, Colo, American Water Works Association.

Moran, M.J., Zogorski, J.S., and Squillace, P.J., 2007, Chlorinated solvents in groundwater of the United States: Environmental Science and Technology, v. 41, no. 1, p. 74–81.

Morrison, P., and Pine, J., 1955, Radiogenic origin of the helium isotopes in rock: Annual New York Academy of Sciences, v. 62, p. 71–92, doi:10.1111/j.1749-6632.1955. tb35366.x, accessed online March, 9, 2011, at http://onlinelibrary.wiley.com/doi/10.1111/j.1749-6632.1955. tb35366.x/abstract.

Nakagaki, N., and Wolock, D.M., 2005, Estimation of agricultural pesticide use in drainage basins using land cover maps and county pesticide data: U.S. Geological Survey Open-File Report 05-1188, 46 p. (Also available at http://pubs.usgs.gov/of/2005/1188/.)

Nriagu, J.O., 1998, History, occurrence, and use of vanadium, in Nriagu, J.O., ed., Vanadium in the environment, Part 1—Chemistry and biochemistry: New York, John Wiley & Sons, Inc., p. 1–24.

Pankow, J.F., Thomson, N.R., Johnson, R.L., Baehr, A.L., and Zogorski, J.S., 1997, The urban atmosphere as a non-point source for the transport of MTBE and other volatile organic compounds (VOCs) to shallow ground water: Environmental Science and Technology, v. 31, no. 10, p. 2821–2828.

Parker, D.R., Seyfferth, A.L., and Kiel Reese, B., 2008, Perchlorate in groundwater, a synoptic survey of "pristine" sites in the coterminous United States: Environmental Science and Technology, v. 42, no. 5, p. 1465–1471.

Piper, A.M., 1944, A graphic procedure in the geochemical interpretation of water analyses: American Geophysical Union Transactions, v. 25, p. 914–923.

Plummer, L.N., Böhlke, J.K., and Doughten, M.W., 2006, Perchlorate in Pleistocene and Holocene groundwater in North-Central New Mexico: Environmental Science and Technology, v. 40, no. 6, p. 1757–1763.

Plummer, L.N., Michel, R.L., Thurman, E.M., and Glynn, P.D., 1993, Environmental tracers for age-dating young ground water, in Alley, W.M., ed., Regional Groundwater Quality: New York, Van Nostrand Reinhold, p. 255–294.

Plummer, L.N., Rupert, M.G., Busenberg, E., and Schlosser, P., 2000, Age of irrigation water in ground water from the Eastern Snake River Plain Aquifer, South-Central Idaho: Ground Water, v. 38, no. 2, p. 264–283.

Price, C.V., Nakagaki, N., Hitt, K.J., and Clawges, R.M., 2003, Mining GIRAS—improving on a national treasure of land use data, in ESRI International Users Conference, 23rd, Redlands, Calif., July 7–11, 2003, Proceedings: Redlands, Calif., ESRI.

Reimann, C., and de Caritat, P., 1998, Chemical elements in the environment: Berlin: Springer-Verlag.

Rowe, B.L., Toccalino, P.L., Moran, M.J., Zogorski, J.S., and Price, C.V., 2007, Occurrence and potential human-health relevance of volatile organic compounds in drinking water from domestic wells in the United States: Environmental Health Perspectives, v. 115, no. 11, p. 1539–1546.

San Diego County Water Authority, 1997, Ground-water report, 147 p.

Saucedo, G.J., 2000, GIS data for the geologic map of California: California Division of Mines and Geology CDMG CD-ROM 2000-007.

Schlosser, P., Stute, M., Dorr, C., Sonntag, C., and Munnich, K.O., 1988, Tritium/^{3}He dating of shallow groundwater: Earth and Planetary Science Letters, v. 89, p. 353–362.

Schlosser, P., Stute, M., Sonntag, C., and Munnich, K.O., 1989, Tritiogenic ^{3}He in shallow groundwater: Earth and Planetary Science Letters, v. 94, p. 245–256.

Scott, J.C., 1990, Computerized stratified random site selection approaches for design of a groundwater quality sampling network: U.S. Geological Survey Water-Resources Investigations Report 90-4101, 109 p. (Also available at http://pubs.er.usgs.gov/djvu/WRI/wrir_90_4101.djvu.)

Solomon, D.K., and Cook, P.G., 2000, ^{3}H and ^{3}He, *in* Cook, P.G., and Herczeg, A.L., eds., Environmental Tracers in Subsurface Hydrology: Boston, Kluwer Academic Press, p. 397–424.

Solomon, D.K., Poreda, R.J., Schiff, S.L., and Cherry, J.A., 1992, Tritium and Helium 3 as groundwater age tracers in the Borden Aquifer: Water Resources Research, v. 28, no. 3, p. 241–755.

Sparks, D.L., 1995, Environmental soil chemistry: San Diego, Academic Press.

State of California, 1999, Supplemental Report of the 1999 Budget Act 1999-00 Fiscal Year, Item 3940-001-0001, State Water Resources Control Board, accessed September 9, 2010, at http://www.lao.ca.gov/1999/99-00_supp_rpt_lang.html#3940.

State of California, 2001a, Assembly Bill No. 599, Chapter 522, accessed September 9, 2010, at http://www.swrcb.ca.gov/gama/docs/ab_599_bill_20011005_chaptered.pdf.

State of California, 2001b, Groundwater Monitoring Act of 2001: California Water Code, part 2.76, Sections 10780-10782.3, accessed September 9, 2010, at http://www.leginfo.ca.gov/cgi-bin/displaycode?section=wat&group=10001-11000&file=10780-10782.3

State of California, 2011, GAMA—Groundwater Ambient Monitoring and Assessment Program: State Water Resources Control Board website, accessed March 9, 2011, at http://www.waterboards.ca.gov/gama/.

State Water Resources Control Board, 2003, A comprehensive groundwater quality monitoring program for California: Assembly Bill 99 Report to the Governor and Legislature, March 2003, 100 p., accessed March 9, 2011, at http://www.waterboards.ca.gov/water_issues/programs/gama/docs/final_ab_599_rpt_to_legis_7_31_03.pdf.

State Water Resources Control Board, 2011: State Water Resources Control Board, Geotracker web page, accessed March 9, 2007, at http://geotracker.swrcb.ca.gov/data_download.asp.

Stollenwerk, K., 2003, Geochemical processes controlling transport of arsenic, in Welch, A.H., and Stollenwerk, K.G., eds., Groundwater—A Review of Adsorption, in Arsenic in Ground Water Geochemistry and Occurrence: Boston, Kluwer Academic Publishers, 475 p.

Thomas, J.M., Welch, A.H., Lico, M.S., Hughes, J.L., and Whitney, R., 1993, Radionuclides in groundwater of the Carson River basin, western Nevada and eastern California, USA: Applied Geochemistry, v. 8, p. 447–471.

Toccalino, P.L., and Norman, J.E., 2006, Health-based screening levels to evaluate U.S. Geological Survey ground water quality data: Risk Analysis, v. 26, no. 5, p. 1339–1348.

Toccalino, P.L., Norman, J.E., Phillips, R.H., Kauffman, L.J., Stackelberg, P.E., Nowell, L.H., Krietzman, S.J., and Post, G.B., 2004, Application of health-based screening levels to groundwater quality data in a state-scale pilot effort: U. S. Geological Survey Scientific Investigations Report Scientific Investigations Report 2004-5174, 64 p. (Also available at http://pubs.usgs.gov/sir/2004/5174.)

Tolstikhin, I.N., and Kamenskiy, I.L., 1969, Determination of groundwater ages by the T-3He method: Geochemistry International, v. 6, p. 310–811.

Torgersen, T., Clarke, W.B., and Jenkins, W.J., 1979, The tritium/helium3 method in hydrology: Vienna, International Atomic Energy Agency, IAEA-SM-228/49, p. 917–930.

Torgersen, T.,1980, Controls on pore-fluid concentrations of ^{4}He and ^{222}Rn and the calculation of ^{4}He/^{222}Rn ages: Journal of Geochemical Exploration, v. 13, p. 7–75.

Torgersen, T., and Clarke, W.B., 1985, Helium accumulation in groundwater—I. An evaluation of sources and continental flux of crustal 4He in the Great Artesian basin, Australia: Geochimica et Cosmochimica Acta, v. 49, p. 1211–1218.

Troiano, J., Weaver, D., Marade, J., Spurlock, F., Pepple, M., Nordmark, C., and Bartkowiak, D., 2001, Summary of well water sampling in California to detect pesticide residues resulting from nonpoint source applications: Journal of Environmental Quality, v. 30, p. 448–459.

U.S. Environmental Protection Agency, 1998, Reporting Requirements for Risk/Benefit Information—Code of Federal Regulations, title 40—Protection of environment, chapter 1—Environmental protection agency, subchapter E—Pesticide programs, part 159—Statements of policies and interpretations, subpart D—reporting requirements for risk/benefit information, 40 CFR 159.184 (National Archives and Records Administration, September 19, 1997; amended June 19, 1998): Federal Register, v. 62, no. 182, accessed September 5, 2008, at http://www.epa.gov/EPA-PEST/1997/September/Day-19/p24937.htm.

U.S. Environmental Protection Agency, 2006, 2006 Edition of the Drinking Water Standards and Health Advisories, updated August 2006: Washington, D.C., U.S. Environmental Protection Agency, Office of Water EPA/822/R-06-013, accessed March 17, 2011, at http://www.epa.gov/waterscience/criteria/drinking/dwstandards.pdf.

U.S. Geological Survey, 2011, What is the Priority Basin Project: California Water Science Center website, accessed March 9, 2011, at http://ca.water.usgs.gov/gama/.

Vine, J.D., and Tourtelot, E.B., 1970, Geochemistry of black shale deposits—a summary report: Economic Geology and the Bulletin of the Society of Economic Geologists, v. 65, no. 3, p. 253–272.

Vogel, J.C., and Ehhalt, D., 1963, The use of the carbon isotopes in groundwater studies, in Conference on Radioisotopes in Hydrology, Tokyo, March 5–9, 1963, Proceedings: Vienna, International Atomic Energy Agency, p. 383–395.

Volgelmann, J.E., Howard, S.M., Yang, L., Larson, C.R., Wylie, B.K., and Van Driel, N., 2001, Completion of the 1990s National Land Cover Data Set for the conterminous Unites States from Landsat Thematic Mapper data and ancillary data sources: Photogrammetric Engineering and Remote Sensing, v. 17, p. 150–612.

Wanty, R.B., and Goldhaber, M.B., 1992, Thermodynamics and kinetics of reactions involving vanadium in natural systems—Accumulation of vanadium in sedimentary rocks: Geochimica et Cosmochimica Acta, v. 56, no. 4, p. 1471–1483.

Wanty, R.B., Goldhaber, M.B., and Northrop, H.R., 1990, Geochemistry of vanadium in an epigenetic, sandstone-hosted vanadium- uranium deposit, Henry Basin, Utah: Economic Geology, v. 85, no. 2, p. 270–284.

Welch, A.H., Lico, M.S., and Hughes, J.L., 1988, Arsenic in ground water of the western United States: Ground Water, v. 26, no. 3, p. 333–347

Welch, A.H., Westjohn, D.B., Helsel, D.R., and Wanty, R.B., 2000, Arsenic in ground water of the United States—occurrence and geochemistry: Ground Water, v. 38, no. 4, p. 589–604.

Welch, A.H., Oremland, R.S., Davis, J.A., and Watkins, S.A., 2006, Arsenic in ground water—a review of current knowledge and relation to the CALFED solution area with recommendations for needed research: San Francisco Estuary and Watershed Science, v. 4, no. 2, Article 4, 32 p., accessed May 19, 2008, at http://repositories.cdlib.org/jmie/sfews/vol4/iss2/art4/.

Wehrli, B., and Stumm, W., 1989, Vanadyl in natural waters—Adsorption and hydrolysis promote oxygenation: Geochimica et Cosmochimica Acta, v. 53, no. 1, p. 69–77.

World Health Organization, 1988, International Programme on Chemical Safety Environmental Health Criteria 81, Vanadium, accessed July 2, 2008, at http://www.inchem.org/documents/ehc/ehc/ehc81 htm.

World Health Organization, 1998, International Programme on Chemical Safety Environmental Health Criteria 204, Boron, accessed January 19, 2010, at http://www.inchem.org/documents/ehc/ehc/ehc204 htm.

Wright, M.T., and Belitz, K., 2010, Factors Controlling the Regional Distribution of Vanadium in Groundwater: Groundwater, v. 48, no. 4, p. 515–525 (also available at http://dx.doi.org/10.1111/j.1745-6584.2009.00666.x.)

Wright, M.T., Belitz, K., and Burton, C.A., 2005, California GAMA program—groundwater quality in the San Diego drainages hydrologic province, California, 2004: U.S. Geological Survey Data Series 129, 91 p. (Also available at http://pubs.usgs.gov/ds/2005/129/.)

Zapecza, O.S., and Szabo, Z., 1988, Natural radioactivity in groundwater—a review: U.S. Geological Survey Water Supply Paper 2325, p. 50–57.

Zogorski, J.S., Carter, J.M., Ivahnenko, T., Lapham, W.W., Moran, M.J., Rowe, B.L., Squillace, P.J., and Toccalino, P.L., 2006, The quality of our Nation's waters—Volatile organic compounds in the Nation's ground water and drinking-water supply wells: U.S. Geological Survey Circular 1292, 101 p. (Also available at http://pubs.usgs.gov/circ/circ1292/.)

Appendix A. Selection of CDHP-Well Data

The strategy used to select CDPH inorganic data for a single well in each cell where the USGS did not obtain a sample for analysis for inorganic constituents involved prioritizing data from different sources. The first choice was to select CDPH data for the grid well sampled by the USGS for other constituents, provided the CDPH data met quality-control criteria and was collected within three years prior to the end of sampling in the San Diego study unit (July 30, 2001 to July 29, 2004). The most recent CDPH data from the well were evaluated to determine whether the cation/ anion balance for the CDPH data less than 10 percent. If so, the CDPH inorganic data from the well were selected for use as grid-well data for inorganic constituents. It was assumed that analyses using major ion data with a cation-anion balance less than 10 percent also resulted in high-quality data for trace elements, nutrients, and radiochemical constituents. This step resulted in the selection of inorganic data from CDPH at 16 wells that also were USGS-grid wells. For identification purposes, data from the CDPH for these grid wells were assigned GAMA identifications numbers equivalent to the GAMA USGS-grid well but with DG inserted between the study area prefix and sequence number (for example, CDPH-grid well SDTEM-DG-08 is the same well as USGS-grid well SDTEM-08) (table A1).

If the first step did not yield CDPH inorganic data for a grid cell, then the second step was to search the CDPH database to identify the highest randomly ranked well within a cell with a cation/anion imbalance less than 10 percent. This step did not result in the selection of any wells.

If no CDPH wells in a grid cell met the charge-balance criteria or if there was insufficient data to evaluate charge balance, then the third choice for the CDPH-grid well was to select the highest randomly ranked CDPH well with any of the targeted inorganic data. This step resulted in selection of seven grid wells from which CDPH inorganic data were used. For identification purposes, data from the CDPH for these grid wells not collocated with USGS-grid wells were assigned GAMA identifications numbers equivalent to the GAMA USGS-grid well for the cell but with DPH inserted between the study area prefix and sequence number (for example, CDPH-grid well SDTEM-DPH-08 in a grid cell with no USGS wells).

Inorganic data from the CDPH database were used at 23 grid wells (table 2). In combination with USGS-grid well inorganic data (19 wells), inorganic data was available for 42 of the 60 grid cells. Analysis of the combined data sets to evaluate the occurrence of relatively high or moderate concentrations was not affected by differences in laboratory reporting levels (LRLs) between GAMA-collected and CDPH data because concentrations greater than one-half of water-quality benchmarks generally were substantially higher than the highest LRLs. The locations, GAMA identification numbers of grid and understanding wells (fig. A1A-A1C), and attributes of CDPH-grid wells are located table A1. Comparisons between USGS-collected and CDPH data are described in appendix D.

Table A1.   Identification and attributes of grid and understanding wells sampled during May 17–July 29, 2004, and grid wells using data for inorganic constituents from the California Department of Public Health (CDPH), San Diego Ground-Water Ambient Monitoring and Assessment (GAMA) study unit, California.

[SDALLV, Alluvial Basins study area; SDALLVU, Alluvial Basins study area understanding well; SDHDRK, Hard Rock study area; SDHDRKU, Hard Rock study area understanding well; SDTEM, Temecula Valley study area; SDTEMFP, Temecula Valley study area flow path well; SDWARN, Warner Valley study area; DG, CDPH data from well sampled by GAMA; DPH, CDPH data from well not sampled by GAMA; ft, feet; m, meter; LSD, land surface datum; USGS, U.S. Geological Survey; PSW, public-supply well; –, no data]

USGS GAMA well identification number	CDPH GAMA well identification number	Well type	Agricultural land use (percent)	Natural land use (percent)	Urban land use (percent)	Construction information			
						Well depth	Top of uppermost open interval	Bottom of lowermost open interval	Length from top of uppermost open interval to bottom of the lowermost open interval
Grid wells									
SDALLV-01	–	PSW	62	37	1	200	100	180	80
SDALLV-02	SDALLV-DG-12	PSW	5	34	60	130	94	117	23
SDALLV-03	–	PSW	0	34	65	606	222	566	344
SDALLV-04	–	PSW	71	19	10	180	80	180	100
SDALLV-05	SDALLV-DG-05	PSW	58	42	0	582	234	513	279
SDALLV-06	–	PSW	2	33	65	200	100	142	42
SDALLV-07[1]	SDALLV-DG-07	PSW	2	84	14	200	39	–	–
SDALLV-08	SDALLV-DG-08	PSW	38	60	2	87	50	78	28
SDALLV-09	–	PSW	0	7	93	810	690	800	110
SDALLV-10	SDALLV-DG-10	PSW	26	35	39	135	65	130	65
SDALLV-11[1]	SDALLV-DG-11	PSW	0	46	54	148	50	148	98
SDALLV-12	–	PSW	0	58	42	230	60	220	160
SDALLV-13[1]	–	PSW	0	94	6	181	96	176	80
SDALLV-14	–	PSW	38	62	0	80	40	80	40
SDALLV-15	–	PSW	2	71	27	107	54	107	53
SDALLV-16[1]	SDALLV-DG-16	PSW	0	97	3	120	48	120	72
SDHDRK-04	–	PSW	0	100	0	315	–	–	–
SDHDRK-05	–	PSW	0	86	14	450	50	450	400
SDHDRK-06	–	PSW	1	45	55	1,000	52	1000	948
SDHDRK-07	–	PSW	6	80	14	400	97	400	303
SDHDRK-08	–	PSW	5	95	0	500	60	500	440
SDHDRK-09	–	PSW	0	100	0	400	75	400	325
SDHDRK-10	SDHDRK-DG-10	PSW	0	100	0	–	–	–	–
SDHDRK-11	SDHDRK-DG-11	PSW	0	100	0	455	20	455	435
SDHDRK-12	–	PSW	0	100	0	186	60	186	126
SDHDRK-13	–	PSW	0	100	0	46	41	–	–
SDTEM-01	–	PSW	57	36	7	1,000	150	1,000	850
SDTEM-03	SDTEM-DG-03	PSW	5	22	74	–	466	909	443
SDTEM-04[1]	SDTEM-DG-04	PSW	72	25	3	252	95	295	200
SDTEM-05	–	PSW	23	51	26	960	200	900	700
SDTEM-06	–	PSW	23	28	49	–	170	470	300
SDTEM-07	–	PSW	17	18	66	307	60	307	247
SDTEM-08	SDTEM-DG-08	PSW	52	44	4	–	114	426	312
SDTEM-09[1]	SDTEM-DG-09	PSW	43	48	9	970	450	950	500
SDTEM-10[1]	–	PSW	0	100	0	250	50	210	160
SDTEM-11	SDTEM-DG-11	PSW	0	51	49	1,000	340	980	640
SDTEM-12[1]	–	PSW	39	58	3	546	96	542	446
SDTEM-13	–	PSW	35	46	19	860	235	860	625
SDWARN-01	–	PSW	0	100	0	473	113	473	360
SDWARN-02[1]	–	PSW	1	99	0	585	100	575	475
SDWARN-03	–	PSW	0	100	0	550	118	550	432
SDWARN-04	–	PSW	18	82	0	438	170	438	268

Table A1.   Identification and attributes of grid and understanding wells sampled during May 17–July 29 2004, and grid wells using data for inorganic constituents from the California Department of Public Health (CDPH), San Diego Ground-Water Ambient Monitoring and Assessment (GAMA) study unit, California.—Continued

[SDALLV, Alluvial Basins study area; SDALLVU, Alluvial Basins study area understanding well; SDHDRK, Hard Rock study area; SDHDRKU, Hard Rock study area understanding well; SDTEM, Temecula Valley study area; SDTEMFP, Temecula Valley study area flow path well; SDWARN, Warner Valley study area; DG, CDPH data from well sampled by GAMA; DPH, CDPH data from well not sampled by GAMA; ft, feet; m, meter; LSD, land surface datum; USGS, U.S. Geological Survey; PSW, public-supply well; –, no data]

USGS GAMA well identification number	CDPH GAMA well identification number	Well type	Agricultural land use (percent)	Natural land use (percent)	Urban land use (percent)	Construction information			
						Well depth	Top of upper-most open interval	Bottom of lowermost open interval	Length from top of uppermost open interval to bottom of the lowermost open interval
Grid wells—Continued									
SDWARN-05	–	PSW	0	100	0	743	130	743	613
SDWARN-06	–	PSW	0	100	0	730	190	730	540
SDWARN-07	–	PSW	0	99	1	295	70	165	95
SDWARN-08	SDWARN-DG-08	PSW	0	100	0	700	280	600	320
SDWARN-09	SDWARN-DG-09	PSW	0	100	0	642	60	642	582
–	SDALLV-DPH-17	PSW	0	88	12	–	–	–	–
–	SDALLV-DPH-18	PSW	27	54	19	–	–	–	–
–	SDALLV-DPH-19	PSW	0	98	2	–	–	–	–
–	SDTEM-DPH-14	PSW	40	26	34	–	–	–	–
–	SDTEM-DPH-15	PSW	49	43	8	–	–	–	–
–	SDTEM-DPH-16	PSW	4	81	15	–	–	–	–
–	SDTEM-DPH-17	PSW	0	98	2	–	–	–	–
Understanding wells									
SDALLVU-01[2]	–	PSW	67	33	0	–	–	–	–
SDHDRKU-01[2]	–	PSW	0	9	91	906	110	906	796
SDHDRKU-02[2]	–	PSW	0	100	0	92	52	92	40
SDHDRKU-03[2]	–	PSW	25	45	30	510	80	510	430
SDTEMFP-01	–	PSW	35	50	16	2,500	234	2,147	1,913
SDTEMFP-02	–	PSW	30	68	3	858	378	838	460
SDTEMFP-03	–	PSW	32	67	1	865	305	845	540
SDTEMFP-04	–	PSW	29	67	4	482	75	465	390
SDTEMFP-05	–	PSW	19	76	5	280	80	270	190
SDTEMFP-06[2]	–	PSW	20	33	47	–	320	1,110	790
SDTEMFP-07[2]	–	PSW	65	21	13	–	270	1,000	730

[1]Well construction information has been updated subsequent to the publication of the San Diego Ground-Water Ambient Monitoring and Assessment (GAMA) study unit data report (Wright and others, 2005).

[2]Well has been reclassified from grid to understanding subsequent to the publication of the San Diego Ground-Water Ambient Monitoring and Assessment (GAMA) study unit data report (Wright and others, 2005).

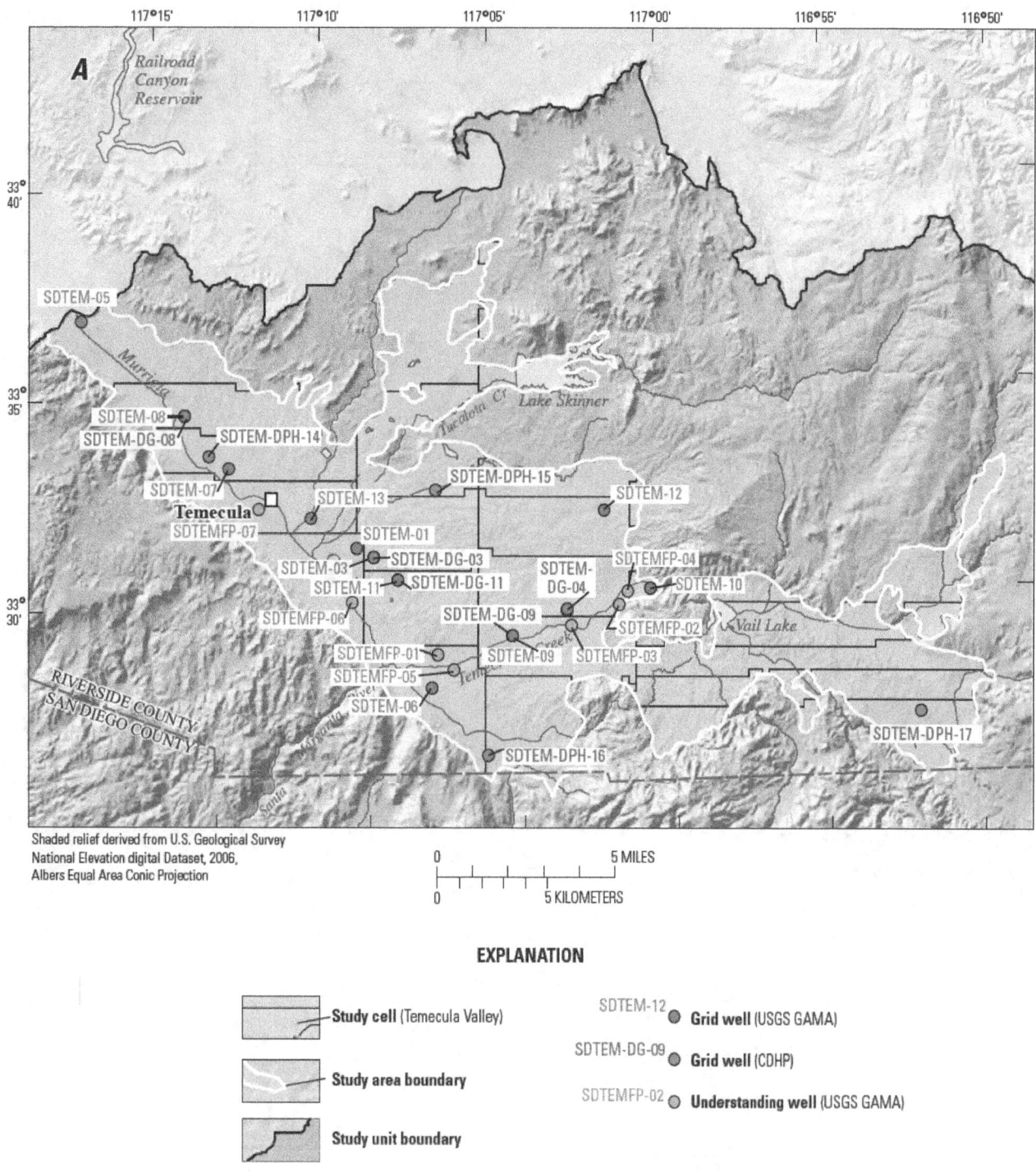

Figure *A1A–A1C*. Identifiers and locations of grid and understanding wells sampled during May–July 2004, and grid wells at which data for inorganic constituents from the California Department of Public Health were used in the study areas of San Diego Groundwater Ambient Monitoring and Assessment (GAMA) study unit, California.

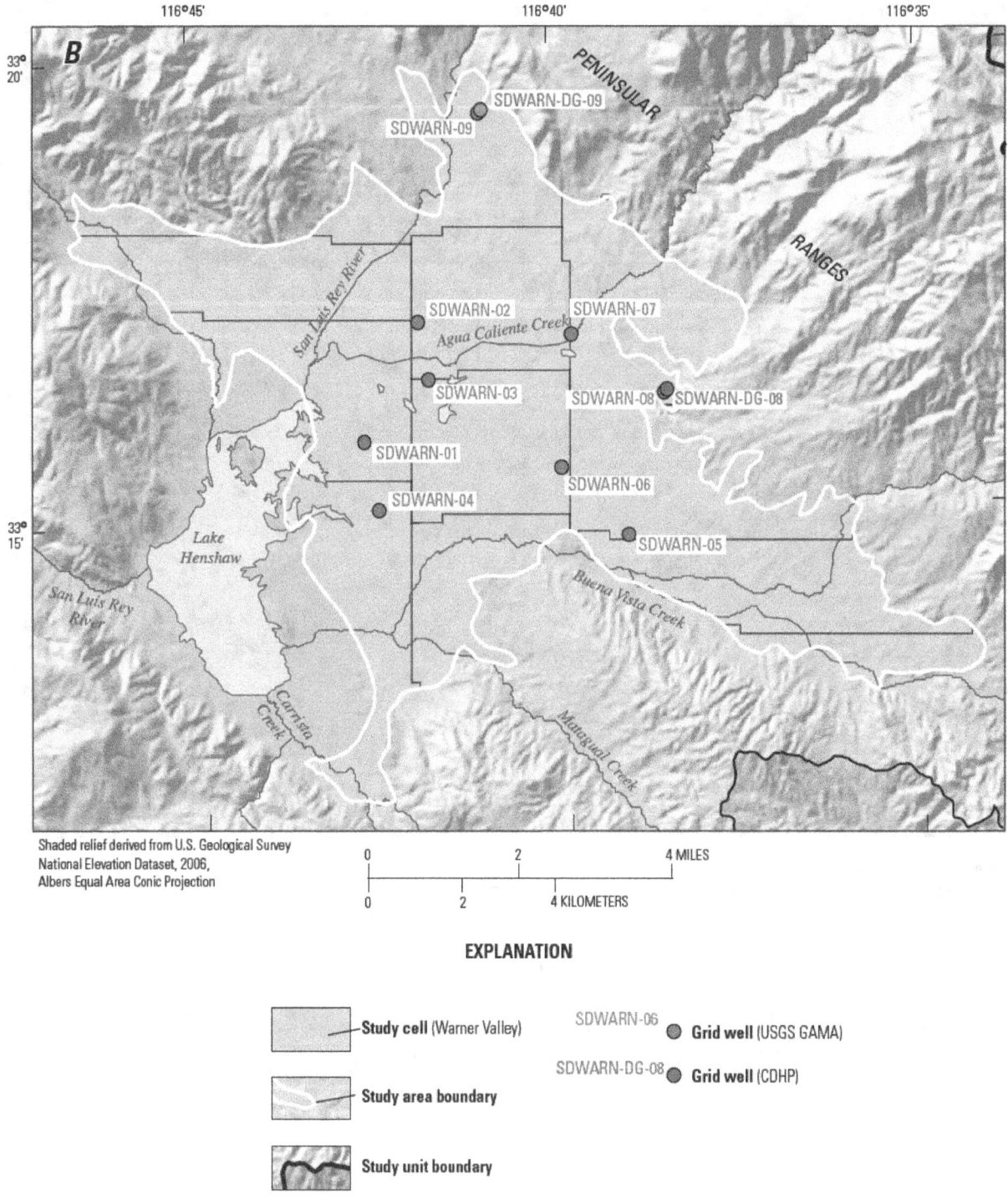

Shaded relief derived from U.S. Geological Survey
National Elevation Dataset, 2006,
Albers Equal Area Conic Projection

EXPLANATION

Study cell (Warner Valley)

Study area boundary

Study unit boundary

SDWARN-06 ● Grid well (USGS GAMA)

SDWARN-DG-08 ● Grid well (CDHP)

Figure *A1A–A1C.*—Continued

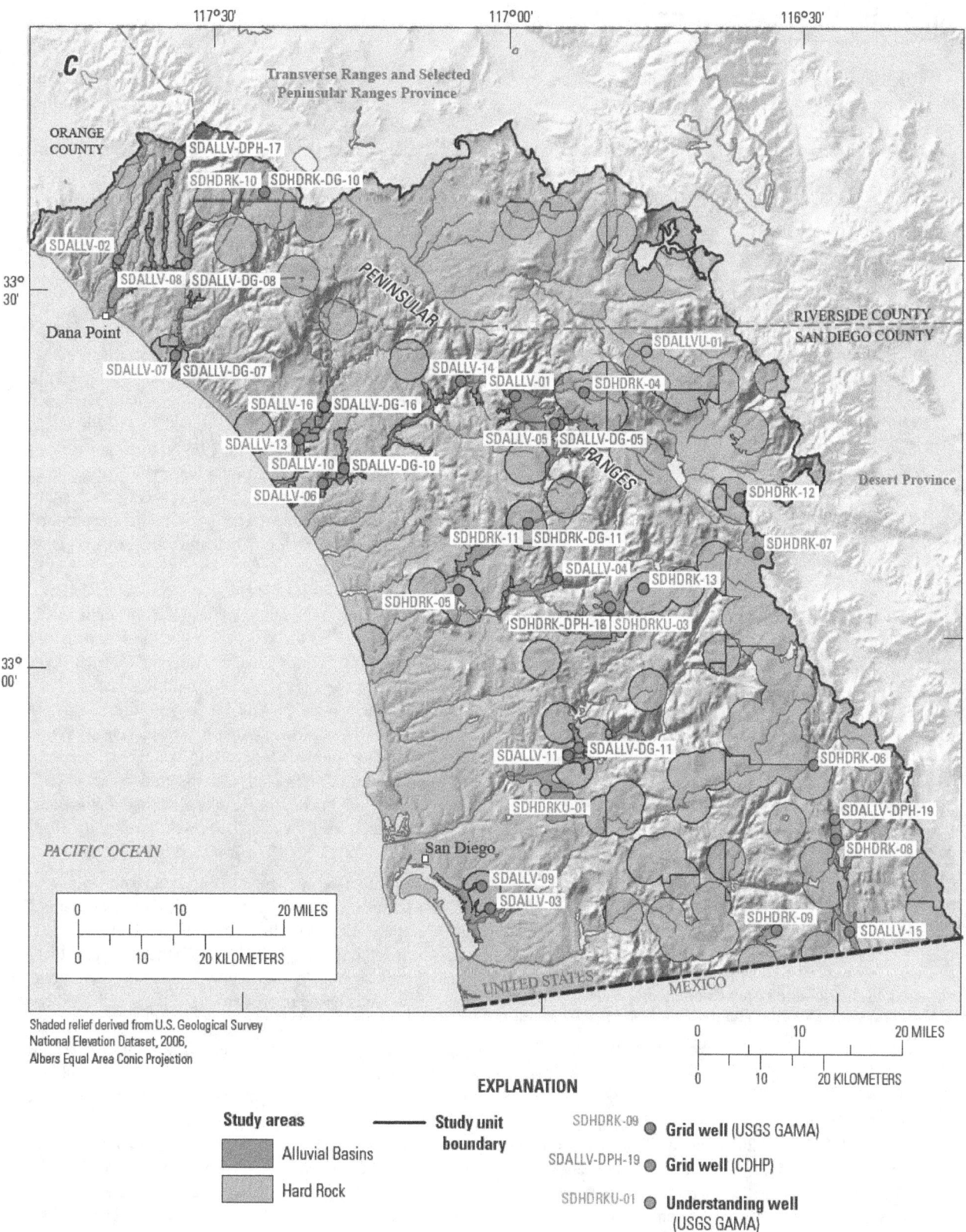

Figure *A1A–A1C.*—Continued

Appendix B. Aquifer Proportion

Three approaches—grid-based, raw detection frequency, and spatially weighted—were selected to evaluate the proportion of the primary supply aquifer in the San Diego study unit with concentrations of constituents greater than water-quality benchmarks (high relative-concentrations).

1. Grid-based: One well in each grid cell was randomly selected to represent the primary aquifers. The relative-concentration for each constituent (concentration relative to its benchmark), or class of constituents, was then evaluated for each grid well. The proportion of the primary aquifers with high relative-concentrations was calculated by taking the number of cells with concentrations greater than the benchmark (relative-concentration > 1), and dividing that number by the total number of grid wells in each of the study areas (Belitz and others, 2010). The proportion for each study area is calculated individually because grid-cell sizes are not uniform across study areas. The proportion for the study unit then is determined by calculating the area-weighted sum using the following equation:

$$AQP_{su} = \sum AQP_{SA} F_{SA,} \qquad (A1)$$

where

AQP_{su} is the aquifer proportion for the study unit,
AQP_{SA} is the aquifer proportion for a study area, and
F_{SA} is the fraction of the total study unit area occupied by the study area.

The F_{SA} Alluvial Fill study areas are: Temecula Valley 0.41, Warner Valley 0.11, and Alluvial Basins 0.48. The proportions of moderate and low relative-concentrations were calculated similarly. Aquifer-scale proportions for individual constituents for each study area (including Hard Rock) are shown in tables B1A-D and for classes of constituents in tables B2A-D. Confidence intervals for grid-based proportions shown in tables B1A-D were computed using the Jeffreys interval for the binomial distribution (Brown and others, 2001). The grid-based estimate is spatially unbiased. However, the grid-based approach may not detect constituents that are present at high relative-concentrations in small proportions of the primary aquifers.

2. Raw detection frequency: Within selected time criteria (July 30 2001–July 29 2004), all available data from the following sources were used to calculate the percentage (frequency) of wells with high relative-concentrations: USGS-grid, CDPH data (most recent analysis per well), and USGS-understanding wells with open intervals representative of the primary aquifers. However, this approach is not spatially unbiased because the CDPH and USGS-understanding wells are not uniformly distributed. Consequently, high relative-concentrations in wells clustering in a particular area represent a small part of the primary aquifers and could be given a disproportionately high weight compared to spatially unbiased methods. Raw detection frequencies of high relative-concentrations are provided for reference in this report but were not used to assess aquifer-scale proportions.

3. Spatially weighted: Similar to the detection frequency approach, the USGS-grid, CDPH data (one analysis per well), and USGS-understanding well data are considered for the spatially weighted approach (Belitz and others, 2010). However for the spatially weighted approach, proportions are computed on a cell-by-cell basis (Isaaks and Srivistava, 1989) rather than as an average of all the wells. Using the spatially weighted approach, the proportion of high values for each study area was computed by (step 1) computing the proportion of high wells in each grid cell and (step 2) averaging the grid-cell values computed in step The spatially weighted high proportions for the study unit was calculated by summing the area-weight values for each study area in the same manner as was used in the grid-based approach. Calculations for individual constituents for each study area are shown in tables B1A-D and for classes of constituents in tables B2A-D. The resulting proportions are spatially unbiased (Isaaks and Srivistava, 1989). Confidence intervals for spatially weighted detection frequencies of high relative-concentrations are not described in this report. Results are based on the most recent analyses for each CDPH well during July 30, 2001–July 29, 2004 (3-year period of study).

The grid-based and spatially weighted estimation of aquifer-scale proportions, based on a spatially distributed grid-cell network across the study unit, are intended to characterize the water-quality of the aquifer at depths that are typically used for public supply. These approaches assign weights to wells based upon a single well per cell (grid-based) or the number of wells per cells (spatially weighted). Another possible approach is to assign weights to wells based on water use (withdrawal rate). However, water use data for public-supply and other wells are not readily available. Moreover, this approach, even if withdrawal data were available for all wells, would characterize the volume of groundwater currently used for public supply, which likely would be weighted towards fewer wells and smaller areas than the approaches used, which were based on spatially distributed grid cells across the study unit.

Table B1A. Current aquifer-scale proportions using grid-based and spatially weighted methods for constituents detected in Temecula Valley study area (1) with concentrations greater than water-quality benchmarks during the period from July 30, 2001 to July 29, 2004 in the California Department of Public Health (CDPH) database, or (2) with high or moderate concentrations in samples collected from grid wells, or (3) with organic constituents detected in more than 10 percent of the grid wells sampled. Grid-based aquifer-scale proportions for organic constituents are based on samples collected by the U.S. Geological Survey from 12 grid wells in the Temecula Valley study area during May–July 2004, San Diego Groundwater Ambient Monitoring and Assessment (GAMA) study unit, California. Spatially weighted aquifer-scale proportions are based on 3 years of CDPH data collected from July 30, 2001–July 29, 2004 in combination with grid and understanding well data.

[High, concentrations greater than water-quality benchmark; moderate, concentrations greater than or equal to 0.1 of benchmark but less than or equal to benchmark for organic constituents (threshold for inorganic constituents is 0.5 of benchmark); low, concentrations less than or equal to 0.1 of benchmark for organic constituents (threshold for inorganic constituents is 0.5 of benchmark); MCL–US; U.S. Environmental Protection Agency maximum contaminant level; MCL–CA, CDPH maximum contaminant level; NL–CA, CDPH notification level; SMCL–CA, CDPH secondary maximum contaminant level; AMCL–US, U.S. Environmental Protection Agency alternate maximum contaminant level; µg/L, micrograms per liter; mg/L, milligrams per liter; pCi/L, picocuries per liter]

Constituent	Threshold type	Threshold value	Threshold units	Grid-based[1]				Raw detection frequency[1]			Spatially weighted[1]		90 percent confidence interval for grid-based high proportion[2]	
				Number of wells	High aquifer proportion (percent)	Moderate aquifer proportion (percent)	Low aquifer proportion (percent)	Number of wells	Number of wells with high values	Raw detection frequency (percent)	Number of cells	High aquifer proportion (percent)	Lower limit (percent)	Upper limit (percent)
Trace elements														
Vanadium	NL–CA	50	µg/L	11	18.2	54.5	27.3	46	11	23.9	13	19.5	5.4	41.9
Arsenic	MCL–US	10	µg/L	10	10.0	0.0	90.0	45	4	8.9	12	9.3	1.8	33.1
Boron	NL–CA	1,000	µg/L	10	10.0	0.0	90.0	42	2	4.8	12	3.1	1.8	33.1
Fluoride	MCL–CA	2	mg/L	10	0.0	0.0	100.0	45	1	2.2	12	1.7	0.0	17.1
Antimony	MCL–US	6	µg/L	10	0.0	0.0	100.0	45	0	0.0	12	0.0	0.0	17.1
Selenium	MCL–US	50	µg/L	10	0.0	0.0	100.0	45	0	0.0	12	0.0	0.0	17.1
Thallium	MCL–US	2	µg/L	10	0.0	0.0	100.0	45	0	0.0	12	0.0	0.0	17.1
Chromium	MCL–CA	50	µg/L	10	0.0	0.0	100.0	46	0	0.0	12	0.0	0.0	17.1
Lead	AL–US	15	µg/L	10	0.0	0.0	100.0	45	0	0.0	12	0.0	0.0	17.1
Nickel	MCL–CA	100	µg/L	10	0.0	0.0	100.0	45	0	0.0	12	0.0	0.0	17.1
Aluminum	MCL–CA	1,000	µg/L	10	0.0	0.0	100.0	45	0	0.0	12	0.0	0.0	17.1
Cadmium	MCL–US	5	µg/L	10	0.0	0.0	100.0	45	0	0.0	12	0.0	0.0	17.1
Mercury	MCL–US	2	µg/L	10	0.0	0.0	100.0	45	0	0.0	12	0.0	0.0	17.1
Radioactive constituents														
Gross-alpha	MCL–US	15	pCi/L	10	0.0	10.0	90.0	46	0	0.0	12	0.0	0.0	17.1
Radon-222	Proposed AMCL–US	4,000	pCi/L	6	0.0	0.0	100.0	11	0	0.0	8	0.0	0.0	26.4
Uranium	MCL–US	30	µg/L	8	0.0	0.0	100.0	18	0	0.0	9	0.0	0.0	20.8
Radium-228	MCL–US	5	pCi/L	6	0.0	0.0	100.0	11	0	0.0	8	0.0	0.0	26.4
Gross-beta radioactivity	MCL–CA	50	pCi/L	6	0.0	0.0	100.0	12	0	0.0	8	0.0	0.0	26.4
Nutrients														
Nitrate, as nitrogen	MCL–US	10	mg/L	12	0.0	8.3	91.7	53	0	0.0	15	0.0	0.0	14.5
Nitrite, as nitrogen	MCL–US	1	mg/L	9	0.0	0.0	100.0	46	0	0.0	12	0.0	0.0	18.7

Table B1A. Current aquifer-scale proportions using grid-based and spatially weighted methods for constituents detected in Temecula Valley study area (1) with concentrations greater than water-quality benchmarks during the period from July 30, 2001 to July 29, 2004 in the California Department of Public Health (CDPH) database, or (2) with high or moderate concentrations in samples collected from grid wells, or (3) with organic constituents detected in more than 10 percent of the grid wells sampled. Grid-based aquifer-scale proportions for organic constituents are based on samples collected by the U.S. Geological Survey from 12 grid wells in the Temecula Valley study area during May–July 2004, San Diego Groundwater Ambient Monitoring and Assessment (GAMA) study unit, California. Spatially weighted aquifer-scale proportions are based on 3 years of CDPH data collected from July 30, 2001–July 29, 2004 in combination with grid and understanding well data.—Continued

[High, concentrations greater than water-quality benchmark; moderate, concentrations greater than or equal to 0.5 of benchmark but less than or equal to 0.1 of benchmark for organic constituents (threshold for inorganic constituents is 0.5 of benchmark); low, concentrations less than or equal to 0.1 of benchmark for organic constituents (threshold for inorganic constituents is 0.5 of benchmark); MCL-US, U.S. Environmental Protection Agency maximum contaminant level; MCL-CA, CDPH maximum contaminant level; NL-CA, CDPH notification level; SMCL-CA, CDPH secondary maximum contaminant level; AMCL-US, U.S. Environmental Protection Agency alternate maximum contaminant level; µg/L, micrograms per liter; mg/L, milligrams per liter; pCi/L, picocuries per liter]

Constituent	Threshold type	Threshold value	Threshold units	Grid-based[1] Number of wells	High aquifer proportion (percent)	Moderate aquifer proportion (percent)	Low aquifer proportion (percent)	Raw detection frequency[1] Number of wells	Number of wells with high values	Raw detection frequency (percent)	Spatially weighted[1] Number of cells	High aquifer proportion (percent)	90 percent confidence interval for grid-based high proportion[2] Lower limit (percent)	Upper limit (percent)
Major and minor elements (SMCLs)														
Total dissolved solids	SMCL-CA	1,000	mg/L	11	0.0	9.1	90.9	47	0	0.0	13	0.0	0.0	15.7
Manganese	SMCL-CA	50	µg/L	10	0.0	0.0	100.0	47	6	12.8	12	7.6	0.0	17.1
Iron	SMCL-CA	300	µg/L	10	0.0	0.0	100.0	47	1	2.1	12	3.0	0.0	17.1
Chloride	SMCL-CA	500	mg/L	10	0.0	0.0	100.0	45	0	0.0	12	0.0	0.0	17.1
Sulfate	SMCL-CA	500	mg/L	10	0.0	0.0	100.0	45	0	0.0	12	0.0	0.0	17.1
Zinc	SMCL-CA	5,000	µg/L	10	0.0	0.0	100.0	44	0	0.0	12	0.0	0.0	17.1
Gasoline components														
Benzene	MCL-CA	1	µg/L	12	0.0	0.0	100.0	42	0	0.0	13	0.0	0.0	14.5
MTBE	MCL-CA	13	µg/L	12	0.0	0.0	100.0	42	0	0.0	13	0.0	0.0	14.5
Trihalomethanes														
Chloroform	MCL-US	[3]380	µg/L	12	0.0	0.0	100.0	42	0	0.0	13	0.0	0.0	14.5
Solvents														
Tetrachloroethylene	MCL-US	5	µg/L	12	0.0	0.0	100.0	42	0	0.0	13	0.0	0.0	14.5
Carbon tetrachloride	MCL-CA	0.5	µg/L	12	0.0	0.0	100.0	42	0	0.0	13	0.0	0.0	14.5
Trichloroethylene	MCL-US	5	µg/L	12	0.0	0.0	100.0	42	0	0.0	13	0.0	0.0	14.5
1,2-Dichloropropane	MCL-US	5	µg/L	12	0.0	0.0	100.0	42	0	0.0	13	0.0	0.0	14.5
Herbicides														
Prometon	HAL-US	100	µg/L	12	0.0	0.0	100.0	12	0	0.0	12	0.0	0.0	14.5
Simazine	MCL-US	4	µg/L	12	0.0	0.0	100.0	39	0	0.0	12	0.0	0.0	14.5
Atrazine	MCL-CA	1	µg/L	12	0.0	0.0	100.0	39	0	0.0	12	0.0	0.0	14.5

Table B1A. Current aquifer-scale proportions using grid-based and spatially weighted methods for constituents detected in Temecula Valley study area (1) with concentrations greater than water-quality benchmarks during the period from July 30, 2001 to July 29, 2004 in the California Department of Public Health (CDPH) database, or (2) with high or moderate concentrations in samples collected from grid wells, or (3) with organic constituents detected in more than 10 percent of the grid wells sampled. Grid-based aquifer-scale proportions for organic constituents are based on samples collected by the U.S. Geological Survey from 12 grid wells in the Temecula Valley study area during May–July 2004, San Diego Groundwater Ambient Monitoring and Assessment (GAMA) study unit, California. Spatially weighted aquifer-scale proportions are based on 3 years of CDPH data collected from July 30, 2001–July 29, 2004 in combination with grid and understanding well data.—Continued

[High, concentrations greater than water-quality benchmark; moderate, concentrations greater than or equal to 0.1 of benchmark but less than or equal to benchmark for organic constituents (threshold for inorganic constituents is 0.5 of benchmark); low, concentrations less than or equal to 0.1 of benchmark for organic constituents (threshold for inorganic constituents is 0.5 of benchmark); MCL-US, U.S. Environmental Protection Agency maximum contaminant level; MCL-CA, CDPH maximum contaminant level; NL-CA, CDPH notification level; SMCL-CA, CDPH secondary maximum contaminant level; AMCL-US, U.S. Environmental Protection Agency alternate maximum contaminant level; µg/L, micrograms per liter; mg/L, milligrams per liter; pCi/L, picocuries per liter]

Constituent	Threshold type	Threshold value	Threshold units	Grid-based[1]				Raw detection frequency[1]			Spatially weighted[1]		90 percent confidence interval for grid-based high proportion[2]	
				Number of wells	High aquifer proportion (percent)	Moderate aquifer proportion (percent)	Low aquifer proportion (percent)	Number of wells	Number of wells with high values	Raw detection frequency (percent)	Number of cells	High aquifer proportion (percent)	Lower limit (percent)	Upper limit (percent)
Constituent of special interest														
Perchlorate	MCL-CA	6	µg/L	7	0.0	66.7	33.3	26	0	0.0	12	0.0	0.0	23.2

[1]Based on most recent analysis for each CDPH well during July 30, 2001–July 29, 2004, combined with GAMA grid-based data.

[2]Based on the Jeffrey's interval for the binomial distribution (Brown and others, 2001).

[3]The MCL-US threshold for trihalomethanes is the sum of chloroform, bromoform, bromodichloromethane, and dibromochloromethane.

Table B1B. Current aquifer-scale proportions using grid-based and spatially weighted methods for constituents detected in Warner Valley study area (1) with concentrations greater than water-quality benchmarks during the period from July 30, 2001 to July 29, 2004 in the California Department of Public Health (CDPH) database, or (2) with high or moderate concentrations in samples collected from grid wells, or (3) with organic constituents detected in more than 10 percent of the grid wells sampled. Grid-based aquifer-scale proportions for organic constituents are based on samples collected by the U.S. Geological Survey from nine grid wells in the Warner Valley study area during May–July 2004, San Diego Groundwater Ambient Monitoring and Assessment (GAMA) study unit, California. Spatially weighted aquifer-scale proportions are based on 3 years of CDPH data collected from July 30, 2001–July 29, 2004 in combination with grid and understanding well data.

[High, concentrations greater than water-quality benchmark; moderate, concentrations greater than or equal to 0.5 of benchmark; low, concentrations less than 0.5 of benchmark); low, concentrations less than or equal to 0.1 of benchmark for organic constituents (threshold for inorganic constituents is 0.5 of benchmark); moderate, concentrations greater than or equal to 0.1 of benchmark but less than or equal to benchmark for organic constituents (threshold for inorganic constituents is 0.5 of benchmark); low, concentrations less than or equal to 0.1 of benchmark for organic constituents (threshold for inorganic constituents is 0.5 of benchmark); MCL-US, U.S. Environmental Protection Agency maximum contaminant level; MCL-CA, CDPH maximum contaminant level; NL-CA, CDPH notification level; SMCL-CA, CDPH secondary maximum contaminant level; AMCL-US, U.S. Environmental Protection Agency alternate maximum contaminant level; µg/L, micrograms per liter; mg/L, milligrams per liter; pCi/L, picocuries per liter]

| Constituent | Threshold type | Threshold value | Threshold units | Grid-based[1] | | | | Raw detection frequency[1] | | | Spatially weighted[1] | | 90 percent confidence interval for grid-based high proportion[2] | |
				Number of wells	High aquifer proportion (percent)	Moderate aquifer proportion (percent)	Low aquifer proportion (percent)	Number of wells	Number of wells with high values	Raw detection frequency (percent)	Number of cells	High aquifer proportion (percent)	Lower limit (percent)	Upper limit (percent)
Trace elements														
Vanadium	NL-CA	50	µg/L	4	0.0	25.0	75.0	5	0	0.0	4	0.0	0.0	36.2
Arsenic	MCL-US	10	µg/L	4	0.0	25.0	75.0	5	0	0.0	4	0.0	0.0	36.2
Fluoride	MCL-CA	2	mg/L	4	0.0	25.0	75.0	5	0	0.0	4	0.0	0.0	36.2
Boron	NL-CA	1,000	µg/L	4	0.0	0.0	100.0	5	0	0.0	4	0.0	0.0	36.2
Antimony	MCL-US	6	µg/L	4	0.0	0.0	100.0	5	0	0.0	4	0.0	0.0	36.2
Selenium	MCL-US	50	µg/L	4	0.0	0.0	100.0	5	0	0.0	4	0.0	0.0	36.2
Thallium	MCL-US	2	µg/L	4	0.0	0.0	100.0	5	0	0.0	4	0.0	0.0	36.2
Chromium	MCL-CA	50	µg/L	4	0.0	0.0	100.0	5	0	0.0	4	0.0	0.0	36.2
Lead	AL-US	15	µg/L	4	0.0	0.0	100.0	5	0	0.0	4	0.0	0.0	36.2
Nickel	MCL-CA	100	µg/L	4	0.0	0.0	100.0	5	0	0.0	4	0.0	0.0	36.2
Aluminum	MCL-CA	1,000	µg/L	4	0.0	0.0	100.0	5	0	0.0	4	0.0	0.0	36.2
Cadmium	MCL-US	5	µg/L	4	0.0	0.0	100.0	5	0	0.0	4	0.0	0.0	36.2
Mercury	MCL-US	2	µg/L	4	0.0	0.0	100.0	5	0	0.0	4	0.0	0.0	36.2
Radioactive constituents														
Radon-222	Proposed AMCL-US	4,000	pCi/L	3	0.0	0.0	100.0	3	0	0.0	3	0.0	0.0	44.4
Gross-alpha	MCL-US	15	pCi/L	4	0.0	0.0	100.0	3	0	0.0	4	0.0	0.0	36.2
Uranium	MCL-US	30	µg/L	3	0.0	0.0	100.0	3	0	0.0	3	0.0	0.0	44.4
Radium-228	MCL-US	5	pCi/L	3	0.0	0.0	100.0	3	0	0.0	3	0.0	0.0	44.4
Gross-beta radioactivity	MCL-CA	50	pCi/L	3	0.0	0.0	100.0	3	0	0.0	3	0.0	0.0	44.4
Nutrients														
Nitrate, as nitrogen	MCL-US	10	mg/L	5	0.0	0.0	100.0	6	0	0.0	5	0.0	0.0	30.6
Nitrite, as nitrogen	MCL-US	1	mg/L	5	0.0	0.0	100.0	6	0	0.0	5	0.0	0.0	30.6

Table B1B. Current aquifer-scale proportions using grid-based and spatially weighted methods for constituents detected in Warner Valley study area (1) with concentrations greater than water-quality benchmarks during the period from July 30, 2001 to July 29, 2004 in the California Department of Public Health (CDPH) database, or (2) with high or moderate concentrations in samples collected from grid wells, or (3) with organic constituents detected in more than 10 percent of the grid wells sampled. Grid-based aquifer-scale proportions for organic constituents are based on samples collected by the U.S. Geological Survey from nine grid wells in the Warner Valley study area during May–July 2004, San Diego Groundwater Ambient Monitoring and Assessment (GAMA) study unit, California. Spatially weighted aquifer-scale proportions are based on 3 years of CDPH data collected from July 30, 2001–July 29, 2004 in combination with grid and understanding well data.—Continued

[High, concentrations greater than water-quality benchmark; moderate, concentrations greater than or equal to 0.1 of benchmark but less than or equal to benchmark for organic constituents (threshold for inorganic constituents is 0.5 of benchmark); low, concentrations less than or equal to 0.1 of benchmark for organic constituents (threshold for inorganic constituents is 0.5 of benchmark); MCL–US; U.S. Environmental Protection Agency maximum contaminant level; MCL–CA, CDPH maximum contaminant level; NL–CA, CDPH notification level; SMCL–CA, CDPH secondary maximum contaminant level; AMCL–US, U.S. Environmental Protection Agency alternate maximum contaminant level; µg/L, micrograms per liter; mg/L, milligrams per liter; pCi/L, picocuries per liter]

| Constituent | Threshold type | Threshold value | Threshold units | Grid-based[1] | | | | Raw detection frequency[1] | | | Spatially weighted[1] | | 90 percent confidence interval for grid-based high proportion[2] | |
				Number of wells	High aquifer proportion (percent)	Moderate aquifer proportion (percent)	Low aquifer proportion (percent)	Number of wells	Number of wells with high values	Raw detection frequency (percent)	Number of cells	High aquifer proportion (percent)	Lower limit (percent)	Upper limit (percent)
Major and minor elements (SMCLs)														
Manganese	SMCL–CA	50	µg/L	4	0.0	0.0	100.0	5	0	0.0	4	0.0	0.0	36.2
Total dissolved solids	SMCL–CA	1,000	mg/L	4	0.0	0.0	100.0	5	0	0.0	4	0.0	0.0	36.2
Iron	SMCL–CA	300	µg/L	4	0.0	0.0	100.0	5	0	0.0	4	0.0	0.0	36.2
Chloride	SMCL–CA	500	mg/L	4	0.0	0.0	100.0	5	0	0.0	4	0.0	0.0	36.2
Sulfate	SMCL–CA	500	mg/L	4	0.0	0.0	100.0	5	0	0.0	4	0.0	0.0	36.2
Zinc	SMCL–CA	5,000	µg/L	4	0.0	0.0	100.0	5	0	0.0	4	0.0	0.0	36.2
Gasoline components														
Benzene	MCL–CA	1	µg/L	9	0.0	0.0	100.0	10	0	0.0	9	0.0	0.0	17.1
MTBE	MCL–CA	13	µg/L	9	0.0	0.0	100.0	10	0	0.0	9	0.0	0.0	17.1
Trihalomethanes														
Chloroform	MCL–US	[3]80	µg/L	9	0.0	0.0	100.0	10	0	0.0	9	0.0	0.0	17.1
Solvents														
Tetrachloroethylene	MCL–US	5	µg/L	9	0.0	0.0	100.0	10	0	0.0	9	0.0	0.0	17.1
Carbon tetrachloride	MCL–CA	0.5	µg/L	9	0.0	0.0	100.0	10	0	0.0	9	0.0	0.0	17.1
Trichloroethylene	MCL–US	5	µg/L	9	0.0	0.0	100.0	10	0	0.0	9	0.0	0.0	17.1
1,2-Dichloropropane	MCL–US	5	µg/L	9	0.0	0.0	100.0	10	0	0.0	9	0.0	0.0	17.1
Herbicides														
Prometon	HAL–US	100	µg/L	9	0.0	0.0	100.0	9	0	0.0	9	0.0	0.0	18.7
Simazine	MCL–US	4	µg/L	9	0.0	0.0	100.0	9	0	0.0	9	0.0	0.0	18.7
Atrazine	MCL–CA	1	µg/L	9	0.0	0.0	100.0	9	0	0.0	9	0.0	0.0	18.7

Table B1B. Current aquifer-scale proportions using grid-based and spatially weighted methods for constituents detected in Warner Valley study area (1) with concentrations greater than water-quality benchmarks during the period from July 30, 2001 to July 29, 2004 in the California Department of Public Health (CDPH) database, or (2) with high or moderate concentrations in samples collected from grid wells, or (3) with organic constituents detected in more than 10 percent of the grid wells sampled. Grid-based aquifer-scale proportions for organic constituents are based on samples collected by the U.S. Geological Survey from nine grid wells in the Warner Valley study area during May–July 2004, San Diego Groundwater Ambient Monitoring and Assessment (GAMA) study unit, California. Spatially weighted aquifer-scale proportions are based on 3 years of CDPH data collected from July 30, 2001–July 29, 2004 in combination with grid and understanding well data.—Continued

[High, concentrations greater than water-quality benchmark; moderate, concentrations greater than or equal to 0.5 of benchmark but less than or equal to benchmark for organic constituents (threshold for inorganic constituents is 0.5 of benchmark); low, concentrations less than or equal to 0.1 of benchmark for organic constituents (threshold for inorganic constituents is 0.5 of benchmark); MCL–US, U.S. Environmental Protection Agency maximum contaminant level; MCL–CA, CDPH maximum contaminant level; NL–CA, CDPH notification level; SMCL–CA, CDPH secondary maximum contaminant level; AMCL–US, U.S. Environmental Protection Agency alternate maximum contaminant level; µg/L, micrograms per liter; mg/L, milligrams per liter; pCi/L, picocuries per liter]

| Constituent | Threshold type | Threshold value | Threshold units | Grid-based[1] | | | | Raw detection frequency[1] | | | Spatially weighted[1] | | 90 percent confidence interval for grid-based high proportion[2] | |
				Number of wells	High aquifer proportion (percent)	Moderate aquifer proportion (percent)	Low aquifer proportion (percent)	Number of wells	Number of wells with high values	Raw detection frequency (percent)	Number of cells	High aquifer proportion (percent)	Lower limit (percent)	Upper limit (percent)
Constituent of special interest														
Perchlorate	MCL–CA	6	µg/L	9	0.0	0.0	100.0	9	0	0.0	9	0.0	0.0	18.7

[1]Based on most recent analysis for each CDPH well during July 30, 2001–July 29, 2004, combined with GAMA grid-based data.

[2]Based on the Jeffrey's interval for the binomial distribution (Brown and others, 2001).

[3]The MCL–US threshold for trihalomethanes is the sum of chloroform, bromoform, bromodichloromethane, and dibromochloromethane.

Table B1C. Current aquifer-scale proportions using grid-based and spatially weighted methods for constituents detected in Alluvial Basins study area (1) with concentrations greater than water-quality benchmarks during the period from July 30, 2001 to July 29, 2004 in the California Department of Public Health (CDPH) database, or (2) with high or moderate concentrations in samples collected from grid wells, or (3) with organic constituents detected in more than 10 percent of the grid wells sampled. Grid-based aquifer-scale proportions for organic constituents are based on samples collected by the U.S. Geological Survey from 16 grid wells in the Alluvial Basins study area during May–July 2004, San Diego Groundwater Ambient Monitoring and Assessment (GAMA) study unit, California. Spatially weighted aquifer-scale proportions are based on 3 years of CDPH data collected from July 30, 2001–July 29, 2004 in combination with grid and understanding well data.

[high, concentrations greater than water-quality benchmark; moderate, concentrations greater than or equal to 0.1 of benchmark but less than or equal to benchmark for organic constituents (threshold for inorganic constituents is 0.5 of benchmark); low, concentrations less than or equal to 0.1 of benchmark for organic constituents (threshold for inorganic constituents is 0.5 of benchmark); MCL-US, U.S. Environmental Protection Agency maximum contaminant level; MCL-CA, CDPH maximum contaminant level; NL-CA, CDPH notification level; SMCL-CA, CDPH secondary maximum contaminant level; AMCL-US, U.S. Environmental Protection Agency alternate maximum contaminant level; µg/L, micrograms per liter; mg/L, milligrams per liter; pCi/L, picocuries per liter]

Constituent	Threshold type	Threshold value	Threshold units	Grid-based Number of wells	High aquifer proportion (percent)	Moderate aquifer proportion (percent)	Low aquifer proportion (percent)	Raw detection Number of wells	Number of wells with high values	Raw detection frequency (percent)	Spatially weighted Number of cells	High aquifer proportion (percent)	Lower limit (percent)	Upper limit (percent)
Trace elements														
Antimony	MCL-US	6	µg/L	14	7.1	0.0	92.9	56	1	1.8	15	6.7	1.3	24.8
Vanadium	NL-CA	50	µg/L	13	0.0	7.1	92.9	66	1	1.5	15	3.3	0.0	13.5
Selenium	MCL-US	50	µg/L	15	0.0	6.7	93.3	67	0	0.0	16	0.0	0.0	11.8
Arsenic	MCL-US	10	µg/L	15	0.0	0.0	100.0	67	0	0.0	16	0.0	0.0	11.8
Boron	NL-CA	1,000	µg/L	14	0.0	0.0	100.0	70	0	0.0	16	0.0	0.0	12.6
Fluoride	MCL-CA	2	mg/L	14	0.0	0.0	100.0	65	0	0.0	15	0.0	0.0	12.6
Thallium	MCL-US	2	µg/L	14	0.0	0.0	100.0	56	0	0.0	16	0.0	0.0	12.6
Chromium	MCL-CA	50	µg/L	14	0.0	0.0	100.0	58	0	0.0	16	0.0	0.0	12.6
Lead	AL-US	15	µg/L	14	0.0	0.0	100.0	64	0	0.0	15	0.0	0.0	12.6
Nickel	MCL-CA	100	µg/L	14	0.0	0.0	100.0	56	0	0.0	16	0.0	0.0	12.6
Aluminum	MCL-CA	1,000	µg/L	15	0.0	0.0	100.0	67	0	0.0	16	0.0	0.0	11.8
Cadmium	MCL-US	5	µg/L	15	0.0	0.0	100.0	67	0	0.0	16	0.0	0.0	11.8
Mercury	MCL-US	2	µg/L	15	0.0	0.0	100.0	67	0	0.0	16	0.0	0.0	11.8
Radioactive constituents														
Gross-alpha	MCL-US	15	pCi/L	15	6.7	13.3	80.0	64	4	6.3	15	7.1	1.2	23.3
Uranium	MCL-US	30	µg/L	11	0.0	27.3	72.7	32	1	3.1	14	1.8	0.0	15.7
Radon-222	Proposed AMCL-US	4,000	pCi/L	6	0.0	0.0	100.0	7	0	0.0	7	0.0	0.0	26.4
Radium-228	MCL-US	5	pCi/L	7	0.0	0.0	100.0	20	0	0.0	9	0.0	0.0	23.2
Gross-beta radioactivity	MCL-CA	50	pCi/L	10	0.0	0.0	100.0	32	0	0.0	11	0.0	0.0	17.1
Nutrients														
Nitrate, as nitrogen	MCL-US	10	mg/L	14	7.1	7.1	85.8	89	7	7.9	16	7.1	1.3	24.8
Nitrite, as nitrogen	MCL-US	1	mg/L	14	0.0	0.0	0.0	69	0	0.0	16	0.0	0.0	12.6

Table B1C. Current aquifer-scale proportions using grid-based and spatially weighted methods for constituents detected in Alluvial Basins study area (1) with concentrations greater than water-quality benchmarks during the period from July 30, 2001 to July 29, 2004 in the California Department of Public Health (CDPH) database, or (2) with high or moderate concentrations in samples collected from grid wells, or (3) with organic constituents detected in more than 10 percent of the grid wells sampled. Grid-based aquifer-scale proportions for organic constituents are based on samples collected by the U.S. Geological Survey from 16 grid wells in the Alluvial Basins study area during May–July 2004, San Diego Groundwater Ambient Monitoring and Assessment (GAMA) study unit, California. Spatially weighted aquifer-scale proportions are based on 3 years of CDPH ata collected from July 30, 2001–July 29, 2004 in combination with grid and understanding well data.—Continued

[high, concentrations greater than water-quality benchmark; moderate, concentrations greater than or equal to benchmark but less than or equal to 0.1 of benchmark for organic constituents (threshold for inorganic constituents is 0.5 of benchmark), low, concentrations less than or equal to 0.1 of benchmark for organic constituents (threshold for inorganic constituents is 0.5 of benchmark), MCL-US; U.S. Environmental Protection Agency maximum contaminant level; MCL-CA, CDPH maximum contaminant level; NL-CA, CDPH notification level; SMCL-CA, CDPH secondary maximum contaminant level; %, percent; mg/L, milligrams per liter; AMCL-US, U.S. Environmental Protection Agency alternate maximum contaminant level; µg/L, micrograms per liter; pCi/L, picocuries per liter]

| Constituent | Threshold type | Threshold value | Threshold units | Grid-based[1] | | | | Raw detection frequency[1] | | | Spatially weighted[1] | | 90 percent confidence interval for grid-based high proportion[2] | |
				Number of wells	High aquifer proportion (percent)	Moderate aquifer proportion (percent)	Low aquifer proportion (percent)	Number of wells	Number of wells with high values	Raw detection frequency (percent)	Number of cells	High aquifer proportion (percent)	Lower limit (percent)	Upper limit (percent)
Major and minor elements (SMCLs)														
Manganese	SMCL-CA	50	µg/L	14	28.6	7.1	64.3	68	26	38.2	16	37.7	12.8	50.3
Total dissolved solids	SMCL-CA	1,000	µg/L	14	28.6	57.1	14.3	67	15	22.4	16	27.1	12.8	50.3
Iron	SMCL-CA	300	µg/L	14	14.3	0.0	85.7	68	9	13.2	16	11.0	4.2	34.2
Chloride	SMCL-CA	500	µg/L	14	7.1	21.4	71.5	67	5	7.5	16	5.7	1.3	24.8
Sulfate	SMCL-CA	500	µg/L	14	7.1	14.3	78.6	67	2	3.0	16	7.8	1.3	24.8
Zinc	SMCL-CA	5,000	µg/L	15	0.0	0.0	100.0	65	0	0.0	16	0.0	0.0	11.8
Gasoline components														
MTBE	MCL-CA	13	µg/L	16	6.3	0.0	93.7	61	2	3.3	17	2.9	1.1	22.0
Benzene	MCL-CA	1	µg/L	16	0.0	0.0	100.0	61	0	0.0	17	0.0	0.0	11.1
Trihalomethanes														
Chloroform	MCL-US	[3]80	µg/L	16	0.0	0.0	100.0	60	0	0.0	16	0.0	0.0	11.1
Solvents														
1,2-Dichloropropane	MCL-US	5	µg/L	16	0.0	6.3	93.7	61	0	0.0	17	0.0	0.0	11.1
Tetrachloroethylene	MCL-US	5	µg/L	16	0.0	0.0	100.0	61	0	0.0	17	0.0	0.0	11.1
Carbon tetrachloride	MCL-CA	0.5	µg/L	16	0.0	0.0	100.0	61	0	0.0	17	0.0	0.0	11.1
Trichloroethylene	MCL-US	5	µg/L	16	0.0	0.0	100.0	61	0	0.0	17	0.0	0.0	11.1
Herbicides														
Prometon	HAL-US	100	µg/L	16	0.0	0.0	100.0	17	0	0.0	16	0.0	0.0	11.1
Simazine	MCL-US	4	µg/L	16	0.0	0.0	100.0	53	0	0.0	16	0.0	0.0	11.1
Atrazine	MCL-CA	1	µg/L	16	0.0	0.0	100.0	53	0	0.0	16	0.0	0.0	11.1

Table B1C. Current aquifer-scale proportions using grid-based and spatially weighted methods for constituents detected in Alluvial Basins study area (1) with concentrations greater than water-quality benchmarks during the period from July 30, 2001 to July 29, 2004 in the California Department of Public Health (CDPH) database, or (2) with high or moderate concentrations in samples collected from grid wells, or (3) with organic constituents detected in more than 10 percent of the grid wells sampled. Grid-based aquifer-scale proportions for organic constituents are based on samples collected by the U.S. Geological Survey from 16 grid wells in the Alluvial Basins study area during May–July 2004, San Diego Groundwater Ambient Monitoring and Assessment (GAMA) study unit, California. Spatially weighted aquifer-scale proportions are based on 3 years of CDPH data collected from July 30, 2001–July 29, 2004 in combination with grid and understanding well data.—Continued

[high, concentrations greater than water-quality benchmark; moderate, concentrations greater than or equal to 0.1 of benchmark but less than or equal to benchmark for organic constituents (threshold for inorganic constituents is 0.5 of benchmark); low, concentrations less than or equal to 0.1 of benchmark for organic constituents (threshold for inorganic constituents is 0.5 of benchmark); MCL-US, U.S. Environmental Protection Agency maximum contaminant level; MCL-CA, CDPH maximum contaminant level; NL-CA, CDPH notification level; SMCL-CA, CDPH secondary maximum contaminant level; AMCL-US, U.S. Environmental Protection Agency alternate maximum contaminant level; μg/L, micrograms per liter; pCi/L, picocuries per liter; %, percent; mg/L, milligrams per liter]

| Constituent | Threshold type | Threshold value | Threshold units | Grid-based[1] | | | | Raw detection frequency[1] | | | Spatially weighted[1] | | | 90 percent confidence interval for grid-based high proportion[2] | |
				Number of wells	High aquifer proportion (percent)	Moderate aquifer proportion (percent)	Low aquifer proportion (percent)	Number of wells	Number of wells with high values	Raw detection frequency (percent)	Number of cells	High aquifer proportion (percent)	Lower limit (percent)	Upper limit (percent)
Constituent of special interest														
Perchlorate	MCL-CA	6	μg/L	16	0.0	18.8	81.2	68	1	1.5	17	0.4	0.0	11.1

[1]Based on most recent analysis for each CDPH well during July 30, 2001–July 29, 2004, combined with GAMA grid-based data.

[2]Based on the Jeffrey's interval for the binomial distribution (Brown and others, 2001).

[3]The MCL–US threshold for trihalomethanes is the sum of chloroform, bromoform, bromodichloromethane, and dibromochloromethane.

Table B1D. Current aquifer-scale proportions using grid-based and spatially weighted methods for constituents detected in Hard Rock study area (1) with concentrations greater than water-quality benchmarks during the period from July 30, 2001 to July 29, 2004 in the California Department of Public Health (CDPH) database, or (2) with high or moderate concentrations in samples collected from grid wells, or (3) with organic constituents detected in more than 10 percent of the grid wells sampled. Grid-based aquifer-scale proportions for organic constituents are based on samples collected by the U.S. Geological Survey from 10 grid wells in the Hard Rock study area during May–July 2004, San Diego Groundwater Ambient Monitoring and Assessment (GAMA) study unit, California. Spatially weighted aquifer-scale proportions are based on 3 years of CDPH data collected from July 30, 2001–July 29, 2004 in combination with grid and understanding well data.

[high, concentrations greater than water-quality benchmark; moderate, concentrations greater than or equal to 0.1 of benchmark but less than or equal to benchmark for organic constituents (threshold for inorganic constituents is 0.5 of benchmark); low, concentrations less than or equal to 0.1 of benchmark for organic constituents (threshold for inorganic constituents is 0.5 of benchmark); MCL-US, U.S. Environmental Protection Agency maximum contaminant level; MCL-CA, CDPH maximum contaminant level; NL-CA, CDPH notification level; SMCL-CA, CDPH secondary maximum contaminant level; AMCL-US, U.S. Environmental Protection Agency alternate maximum contaminant level; μg/L, micrograms per liter; mg/L, milligrams per liter; pCi/L, picocuries per liter]

Constituent	Threshold type	Threshold value	Threshold units	Grid-based[1] Number of wells	High aquifer proportion (percent)	Moderate aquifer proportion (percent)	Low aquifer proportion (percent)	Raw detection frequency[1] Number of wells	Number of wells with high values	Raw detection frequency (percent)	Spatially weighted[1] Number of cells	High aquifer proportion (percent)	90 percent confidence interval for grid-based high proportion[2] Lower limit (percent)	Upper limit (percent)
Trace elements														
Arsenic	MCL-US	10	μg/L	5	0.0	20.0	80.0	45	0	0.0	9	0.0	0.0	30.6
Fluoride	MCL-CA	2	mg/L	5	0.0	0.0	100.0	50	2	4.0	9	2.2	0.0	30.6
Aluminum	MCL-CA	1,000	μg/L	5	0.0	0.0	100.0	47	1	2.1	9	1.2	0.0	30.6
Vanadium	NL-CA	50	μg/L	5	0.0	0.0	100.0	34	0	0.0	9	0.0	0.0	30.6
Boron	NL-CA	1,000	μg/L	5	0.0	0.0	100.0	30	0	0.0	9	0.0	0.0	30.6
Antimony	MCL-US	6	μg/L	5	0.0	0.0	100.0	41	0	0.0	9	0.0	0.0	30.6
Selenium	MCL-US	50	μg/L	5	0.0	0.0	100.0	46	0	0.0	9	0.0	0.0	30.6
Thallium	MCL-US	2	μg/L	5	0.0	0.0	100.0	39	0	0.0	9	0.0	0.0	30.6
Chromium	MCL-CA	50	μg/L	5	0.0	0.0	100.0	42	0	0.0	9	0.0	0.0	30.6
Lead	AL-US	15	μg/L	5	0.0	0.0	100.0	45	0	0.0	9	0.0	0.0	30.6
Nickel	MCL-CA	100	μg/L	5	0.0	0.0	100.0	39	0	0.0	9	0.0	0.0	30.6
Cadmium	MCL-US	5	μg/L	5	0.0	0.0	100.0	46	0	0.0	9	0.0	0.0	30.6
Mercury	MCL-US	2	μg/L	5	0.0	0.0	100.0	46	0	0.0	9	0.0	0.0	30.6
Radioactive constituents														
Radon-222	Proposed AMCL-US	4,000	pCi/L	4	25.0	0.0	75.0	4	1	25.0	4	25.0	4.6	65.1
Gross-alpha	MCL-US	15	pCi/L	6	0.0	50.0	50.0	42	7	16.7	9	18.5	0.0	26.4
Uranium	MCL-US	30	μg/L	5	0.0	20.0	80.0	23	3	13.0	7	7.3	0.0	30.6
Radium-228	MCL-US	5	pCi/L	5	0.0	0.0	100.0	19	2	10.5	7	9.5	0.0	30.6
Gross-beta radioactivity	MCL-CA	50	pCi/L	4	0.0	0.0	100.0	17	0	0.0	8	0.0	0.0	36.2
Nutrients														
Nitrate, as nitrogen	MCL-US	10	mg/L	6	0.0	0.0	100.0	99	1	1.0	10	1.2	0.0	26.4
Nitrite, as nitrogen	MCL-US	1	mg/L	6	0.0	0.0	100.0	69	0	0.0	10	0.0	0.0	26.4

Table B1D. Current aquifer-scale proportions using grid-based and spatially weighted methods for constituents detected in Hard Rock study area (1) with concentrations greater than water-quality benchmarks during the period from July 30, 2001 to July 29, 2004 in the California Department of Public Health (CDPH) database, or (2) with high or moderate concentrations in samples collected from grid wells, or (3) with organic constituents detected in more than 10 percent of the grid wells sampled. Grid-based aquifer-scale proportions for organic constituents are based on samples collected by the U.S. Geological Survey from 10 grid wells in the Hard Rock study area during May–July 2004, San Diego Groundwater Ambient Monitoring and Assessment (GAMA) study unit, California. Spatially weighted aquifer-scale proportions are based on 3 years of CDPH data collected from July 30, 2001–July 29, 2004 in combination with grid and understanding well data.—Continued

[high, concentrations greater than water-quality benchmark; moderate, concentrations greater than or equal to 0.1 of benchmark but less than or equal to benchmark for organic constituents (threshold for inorganic constituents is 0.5 of benchmark); low, concentrations less than or equal to 0.1 of benchmark for organic constituents (threshold for inorganic constituents is 0.5 of benchmark); MCL-US, U.S. Environmental Protection Agency maximum contaminant level; MCL-CA, CDPH maximum contaminant level; NL-CA, CDPH notification level; SMCL-CA, CDPH secondary maximum contaminant level; AMCL-US, U.S. Environmental Protection Agency alternate maximum contaminant level; µg/L, micrograms per liter; pCi/L, picocuries per liter; %, percent; mg/L, milligrams per liter; AMCL-US, U.S. Environmental Protection Agency alternate maximum contaminant level]

Constituent	Threshold type	Threshold value	Threshold units	Grid-based Number of wells	High aquifer proportion (percent)	Moderate aquifer proportion (percent)	Low aquifer proportion (percent)	Raw detection frequency Number of wells	Number of wells with high values	Raw detection frequency (percent)	Spatially weighted Number of cells	High aquifer proportion (percent)	90 percent confidence interval Lower limit (percent)	Upper limit (percent)
Major and minor elements (SMCLs)														
Manganese	SMCL-CA	50	µg/L	6	33.3	16.7	50.0	48	15	31.3	9	27.5	10.4	65.9
Total dissolved solids	SMCL-CA	1,000	mg/L	6	16.7	0.0	83.3	45	2	4.4	10	10.0	3.0	49.5
Iron	SMCL-CA	300	µg/L	6	0.0	33.0	67.0	50	14	28.0	9	29.9	0.0	26.4
Sulfate	SMCL-CA	500	mg/L	5	0.0	20.0	80.0	45	1	2.2	9	1.6	0.0	30.6
Chloride	SMCL-CA	500	mg/L	5	0.0	0.0	100.0	46	1	2.2	10	5.0	0.0	30.6
Zinc	SMCL-CA	5,000	µg/L	5	0.0	0.0	100.0	44	1	2.3	10	1.4	0.0	30.6
Gasoline components														
Benzene	MCL-CA	1	µg/L	10	0.0	0.0	100.0	40	1	2.5	10	1.1	0.0	17.1
MTBE	MCL-CA	13	µg/L	10	0.0	0.0	100.0	46	0	0.0	10	0.0	0.0	17.1
Trihalomethanes														
Chloroform	MCL-US	[3]80	µg/L	8	0.0	0.0	100.0	38	0	0.0	9	0.0	0.0	20.8
Solvents														
Tetrachloroethylene	MCL-US	5	µg/L	10	0.0	0.0	100.0	40	0	0.0	10	0.0	0.0	17.1
Carbon tetrachloride	MCL-CA	0.5	µg/L	10	0.0	0.0	100.0	40	0	0.0	10	0.0	0.0	17.1
Trichloroethylene	MCL-US	5	µg/L	10	0.0	0.0	100.0	40	0	0.0	10	0.0	0.0	17.1
1,2-Dichloropropane	MCL-US	5	µg/L	10	0.0	0.0	100.0	40	0	0.0	10	0.0	0.0	17.1
Herbicides														
Prometon	HAL-US	100	µg/L	10	0.0	0.0	100.0	11	0	0.0	10	0.0	0.0	17.1
Simazine	MCL-US	4	µg/L	10	0.0	0.0	100.0	36	0	0.0	10	0.0	0.0	17.1
Atrazine	MCL-CA	1	µg/L	10	0.0	0.0	100.0	36	0	0.0	10	0.0	0.0	17.1

Table B1D.   Current aquifer-scale proportions using grid-based and spatially weighted methods for constituents detected in Hard Rock study area (1) with concentrations greater than water-quality benchmarks during the period from July 30, 2001 to July 29, 2004 in the California Department of Public Health (CDPH) database, or (2) with high or moderate concentrations in samples collected from grid wells, or (3) with organic constituents detected in more than 10 percent of the grid wells sampled. Grid-based aquifer-scale proportions for organic constituents are based on samples collected by the U.S. Geological Survey from 10 grid wells in the Hard Rock study area during May–July 2004, San Diego Groundwater Ambient Monitoring and Assessment (GAMA) study unit, California. Spatially weighted aquifer-scale proportions are based on 3 years of CDPH data collected from July 30, 2001–July 29, 2004 in combination with grid and understanding well data.—Continued

[high, concentrations greater than water-quality benchmark; moderate, concentrations greater than or equal to 0.1 of benchmark but less than or equal to benchmark for organic constituents (threshold for inorganic constituents is 0.5 of benchmark); low, concentrations less than or equal to 0.1 of benchmark for organic constituents (threshold for inorganic constituents is 0.5 of benchmark); MCL-US, U.S. Environmental Protection Agency maximum contaminant level; MCL-CA, CDPH maximum contaminant level; NL-CA, CDPH notification level; SMCL-CA, CDPH secondary maximum contaminant level; %, percent; mg/L, milligrams per liter; AMCL-US, U.S. Environmental Protection Agency alternate maximum contaminant level; µg/L, micrograms per liter; pCi/L, picocuries per liter]

| Constituent | Threshold type | Threshold value | Threshold units | Grid-based[1] | | | | Raw detection frequency[1] | | | Spatially weighted[1] | | 90 percent confidence interval for grid-based high proportion[2] | |
				Number of wells	High aquifer proportion (percent)	Moderate aquifer proportion (percent)	Low aquifer proportion (percent)	Number of wells	Number of wells with high values	Raw detection frequency (percent)	Number of cells	High aquifer proportion (percent)	Lower limit (percent)	Upper limit (percent)
Constituent of special interest														
Perchlorate	MCL-CA	6	µg/L	10	0.0	0.0	100.0	21	0	0.0	10	0.0	0.0	17.1

[2]Based on most recent analysis for each CDPH well during July 30, 2001–July 29, 2004, combined with GAMA grid-based data.

[1]Based on the Jeffreys interval for the binomial distribution (Brown and others, 2001).

[3]The MCL-US threshold for trihalomethanes is the sum of chloroform, bromoform, bromodichloromethane, and dibromochloromethane.

Table B2A. Grid-based aquifer-scale proportions for constituent classes, Temecula Valley study area, San Diego Groundwater Ambient Monitoring and Assessment (GAMA) Program study unit, California.

[SMCL, secondary maximum contaminant level; values are grid based unless otherwise noted]

Constituent class	Aquifer proportion values (percent)		
	High	Moderate	Low
Inorganics with health-based benchmarks			
Trace elements	27.3	45.5	27.3
Radioactive	0.0	10.0	90.0
Nutrients	0.0	8.3	91.7
Any inorganic with health-based benchmarks	27.3	50.0	22.7
Inorganics with aesthetic benchmarks			
Total dissolved solids and (or) chloride and (or) sulfate	0.0	10.0	90.0
Manganese and (or) iron	[1]7.6	0.0	100.0
Organics with health-based benchmarks			
Trihalomethanes	0.0	0.0	100.0
Solvents	0.0	0.0	100.0
Gasoline components	0.0	0.0	100.0
Pesticides	0.0	0.0	100.0
Any organic with health-based benchmarks	0.0	0.0	100.0
Constituent of special interest			
Perchlorate	0.0	66.7	33.3

[1]Spatially weighted value. Aquifer-scale proportions will not sum to 100 if a spatially weighted value is used.

Table B2B. Grid-based aquifer-scale proportions for constituent classes, Warner Valley study area, San Diego Groundwater Ambient Monitoring and Assessment (GAMA) Program study unit, California.

[SMCL, secondary maximum contaminant level; values are grid based unless otherwise noted]

Constituent class	Aquifer proportion values (percent)		
	High	Moderate	Low
Inorganics with health-based benchmarks			
Trace elements	0.0	20.0	80.0
Radioactive	0.0	0.0	100.0
Nutrients	0.0	0.0	100.0
Any inorganic with health-based benchmarks	0.0	20.0	80.0
Inorganics with aesthetic benchmarks			
Total dissolved solids and (or) chloride and (or) sulfate	0.0	0.0	100.0
Manganese and (or) iron	0.0	0.0	100.0
Organics with health-based benchmarks			
Trihalomethanes	0.0	0.0	100.0
Solvents	0.0	0.0	100.0
Gasoline components	0.0	0.0	100.0
Pesticides	0.0	0.0	100.0
Any organic with health-based benchmarks	0.0	0.0	100.0
Constituent of special interest			
Perchlorate	0.0	0.0	100.0

Table B2C.   Grid-based aquifer-scale proportions for constituent classes, Alluvial Basins study area, San Diego Groundwater Ambient Monitoring and Assessment (GAMA) Program study unit, California.

[SMCL, secondary maximum contaminant level; values are grid based unless otherwise noted]

Constituent class	Aquifer proportion values (percent)		
	High	Moderate	Low
Inorganics with health-based benchmarks			
Trace elements	6.7	13.3	80.0
Radioactive	6.7	13.3	80.0
Nutrients	7.1	7.1	85.8
Any inorganic with health-based benchmarks	13.3	20.0	66.7
Inorganics with aesthetic benchmarks			
Total dissolved solids and (or) chloride and (or) sulfate	28.6	57.1	14.3
Manganese and (or) iron	28.6	7.1	64.3
Organics with health-based benchmarks			
Trihalomethanes	0.0	0.0	100.0
Solvents	0.0	6.3	93.7
Gasoline components	6.3	0.0	93.7
Pesticides	0.0	0.0	100.0
Any organic with health-based benchmarks	6.3	6.3	87.4
Constituents of special interest			
Perchlorate	[1]0.4	18.8	81.2

[1]Spatially weighted value. Aquifer-scale proportions will not sum to 100 if a spatially weighted value is used.

Table B2D.   Grid-based aquifer-scale proportions for constituent classes, Hard Rock study area, San Diego Groundwater Ambient Monitoring and Assessment (GAMA) Program study unit, California.

[SMCL, secondary maximum contaminant level; values are grid based unless otherwise noted]

Constituent class	Aquifer proportion values (percent)		
	High	Moderate	Low
Inorganics with health-based benchmarks			
Trace elements	[1]1.2	20.0	80.0
Radioactive	25.0	25.0	50.0
Nutrients	[1]1.2	0.0	100.0
Any inorganic with health-based benchmarks	25.0	25.0	50.0
Inorganics with aesthetic benchmarks			
Total dissolved solids and (or) chloride and (or) sulfate	16.7	0.0	83.3
Manganese and (or) iron	33.3	16.7	50.0
Organics with health-based benchmarks			
Trihalomethanes	0.0	0.0	100.0
Solvents	0.0	0.0	100.0
Gasoline components	[1]1.1	0.0	100.0
Pesticides	0.0	0.0	100.0
Any organic with health-based benchmarks	0.0	0.0	100.0
Constituents of special interest			
Perchlorate	0.0	0.0	100.0

[1]Spatially weighted value. Aquifer-scale proportions will not sum to 100 if a spatially weighted value is used.

Appendix C. Ancillary Datasets

Land-use classifications and percentages, well construction information, groundwater age data and classifications, and redox classifications are listed in tables A1, C1, and C2.

Land-Use Classification

Land use was classified using an "enhanced" version of the satellite derived (30 m pixel resolution), nationwide USGS National Land Cover Dataset (Volgeman and others, 2001; Price and others, 2003). This dataset has been used in previous national and regional studies relating land use to water quality (Gilliom and others, 2006; Zogorski and others, 2006). The data represent land use during approximately the early 1990s. The imagery is classified into 25 land-cover classifications (Nakagaki and Wolock, 2005). These 25 land-cover classifications were condensed into 3 principal land-use categories: urban, agricultural, and natural. Land-use statistics for the study unit, study areas, and for circles with a radius of 500-m around each study well were calculated for classified datasets using ArcGIS (Johnson and Belitz, 2009).

Well Construction Information

Well construction data primarily were determined from driller's logs. On occasion, well construction data was obtained from ancillary records of well owners or from the USGS National Water Information System database. Well identification verification procedures are described by Wright and others (2005).

Groundwater Age Classification

Groundwater dating techniques provide a measure of the time since the groundwater was last in contact with the atmosphere. Techniques aimed at estimating groundwater residence times or 'age' include those based on tritium (for example, Tolstikhin and Kamensskiy, 1969; Torgersen and others, 1979) and tritium in combination with its decay product helium-3 (Schlosser and others, 1988, 1989; Solomon and others, 1992), carbon-14 activities (for example, Vogel and Ehhalt, 1963; Plummer and others, 1993; Kalin, 2000), and dissolved noble gases, particularly helium-4 accumulation (for example, Andrews and Lee, 1979; Davis and DeWiest, 1966; Kulongoski and others, 2008).

Tritium (^{3}H) is a short-lived radioactive isotope of hydrogen with a half-life of 12.32 years (Lucas and Unterweger, 2000). ^{3}H is produced naturally in the atmosphere by the interaction of cosmogenic radiation with nitrogen, by above-ground nuclear explosions, and by the operation of nuclear reactors. Tritium enters the hydrological cycle following oxidation to tritiated water. Consequently, the presence of ^{3}H in groundwater may be used to identify water that has exchanged with the atmosphere in the past 50 years. By determining the ratio of ^{3}H to ^{3}He, resulting from the radioactive decay of ^{3}H, the time that the water has resided in the aquifer can be calculated more precisely than using tritium alone for water (Solomon and others, 1992)).

The widely used carbon-14 chronometer relies on evaluating the radiocarbon content of dissolved inorganic carbonate species in groundwater. ^{14}C is formed in the atmosphere by the interaction of cosmic-ray neutrons with nitrogen, and to a lesser degree with oxygen and carbon. ^{14}C is incorporated into carbon dioxide and mixed throughout the atmosphere, dissolving in precipitation and entering the hydrologic cycle. ^{14}C activity in groundwater, expressed as percent modern carbon (pmc), indicates exposure to the atmospheric ^{14}C source, and is governed by the decay constant of ^{14}C (with a half-life of 5,730 yrs). ^{14}C can be used to estimate groundwater ages ranging from 1,000 to less than 30,000 years before present because of its half-life. Calculated ^{14}C ages in this study are referred to as "uncorrected" because they have not been adjusted to consider exchanges with sedimentary sources of carbon (Fontes and Garnier, 1979;). The ^{14}C age (residence time) is calculated based on the decrease in ^{14}C activity as a result of radioactive decay with time since groundwater recharge, relative to an assumed initial ^{14}C concentration (Clarke and Fritz, 1997). A mean initial ^{14}C activity of 99 pmc is assumed for this study, with estimated errors on calculated groundwater ages up to 20 percent.

Helium (He) is a naturally occurring inert gas initially included during the accretion of the planet, and later produced by the radioactive decay of lithium, thorium, and uranium in the Earth. Measured groundwater He concentrations represent the sum of several He components including air-equilibrated He (He_{eq}), dissolved-air bubbles (He_a), terrigenic He (He_{ter}), and tritiogenic He-3 (^3He_t). Helium (^{3}He and ^{4}He) concentrations in groundwater often exceed the expected solubility equilibrium values, a function of the temperature of the water, as a result of subsurface production of both isotopes and their subsequent release into the groundwater (for example, Morrison and Pine, 1955; Andrews and Lee, 1979; Torgersen, 1980; Andrews, 1985; Torgersen and Clarke, 1985). The presence of terrigenic He in groundwater, from its production in aquifer material or deeper in the crust, is indicative of long groundwater residence times. The amount of terrigenic helium is defined as the concentration of the total measured helium minus the fraction resulting from air-equilibration [He_{eq}] and dissolved air-bubbles [He_a]. For the purposes of this study, percent terrigenic He is used to identify groundwater with residence times greater than 100 yr. Percent terrigenic He is defined as the concentration of terrigenic He (as defined previously) divided by the total measured He in the sample (corrected for air-bubble entrainment). Samples with greater than, or equal to, 5 percent terrigenic He represent groundwater with a residence time of more than 100 yrs.

Recharge temperatures were calculated from dissolved neon, argon, krypton, and xenon using methods described in Aeschbach-Hertig and others (1999). Only modeled recharge temperatures having a probability greater than 1 percent were accepted (Aeschbach-Hertig and others, 2000), and the sample with the highest probability was used in this report.

^{3}H /^{3}He apparent ages were computed as described in Solomon and Cook (2000). The uncertainty for computed ^{3}H/^{3}He apparent ages is greater in samples with terrigenic He greater than 5 percent because of sensitivity to the ^{3}He/^{4}He ratio of the terrigenic He (Plummer and others, 2000). The ^{3}He/^{4}He ratio of samples was determined by linear regression of the percent of terrigenic He against the δ^3He ([δ^3He = R_{meas}/R_{atm} −1] × 100) of samples with less than 1 tritium unit (TU).

In this study, the age distributions of samples are classified as pre-modern, modern, and mixed. Groundwater with tritium activity less than 1 TU, percent terrigenic He greater than, or equal to, 5 percent, and ^{14}C less than 90 pmc is designated as pre-modern—defined as having recharged prior to 1950. Groundwater with tritium activities greater than 1 TU, percent terrigenic He less than 5 percent, and ^{14}C greater than 90 pmc is designated as modern—defined as having recharged during the last 50 years. Samples with both pre-modern and modern components are designated as mixed groundwater, which includes substantial fractions of both old and young waters. In reality, pre-modern groundwater could contain very small fractions of modern water and modern groundwater could contain small fractions of pre-modern water. Previous investigations have used a range of tritium values from 0.3 to 1.0 TU as thresholds for distinguishing pre-1950 from post-1950 water (Michel, 1989; Plummer and others, 1993; Michel and Schroeder, 1994; Clark and Fritz, 1997; Manning and others, 2005). By using a tritium value of 1.0 TU, at the upper end of the range used in the literature, for the threshold in this study, the age classification scheme allows a slightly larger fraction of modern water to be present in a classified pre-modern age distribution than if a lower threshold were used. A lower threshold for tritium would result in fewer wells classified as having a pre-modern rather than a mixed age distribution, when other tracers, carbon-14 and terrigenic helium, suggested that they were primarily pre-modern water. This higher threshold was considered more appropriate for this study because many of the wells are long-screened production wells and some mixing of at least some waters of different ages is likely to occur.

Tritium, pmc, and percent terrigenic helium, along with sample age classifications are reported in appendix table C1. Because of uncertainties in age distributions, in particular caused by mixing of waters of different ages in wells with long open interval intervals and high withdrawal rates, these more precise age estimates were not specifically used for quantifying the relation between age and water quality in this report. Although more sophisticated lumped parameter models for analyzing age distributions that incorporate mixing are available (Cook and Bohlke, 2000), use of these alternative models to understand age mixtures was beyond the scope of this report. Rather, classification into modern, mixed, and pre-modern categories was considered to provide an appropriate and useful characterization for the purposes of examining groundwater quality.

Geochemical Conditions

Geochemical conditions investigated as potential explanatory variables in this report include oxidation-reduction (redox) characteristics. Microorganisms affect the redox conditions of groundwater by utilizing terminal electron acceptors during the degradation of organic carbon. The order of terminal electron acceptor utilization is: O_2 > NO_3- > Mn (IV) > Fe (III) > SO_4^{2-} > CO_2 (McMahon and Chapelle, 2008). With the successive utilization and subsequent depletion of terminal electron acceptors, the redox condition of groundwater progresses from oxidizing (positive Eh values) to reducing (negative Eh values). Oxidation-reduction (redox) conditions affect the mobility of many organic and inorganic constituents (McMahon and Chapelle, 2008), and thus constituent concentrations were compared to redox conditions of the system.

Classification of redox conditions was done using a modification of the framework of McMahon and Chapelle (2008), and is shown in table C2. Redox conditions were classified as either oxic or anoxic. This study utilizes both filtered (USGS data) and unfiltered (CDPH data) samples to infer redox conditions of groundwater. For filtered samples, concentration thresholds as outlined by McMahon and Chapelle (2008) were used. However, because unfiltered samples may overestimate the concentration of dissolved constituents (see discussion in appendix D), a second set of raised concentration thresholds were used for CDPH data. The difference in concentration between unfiltered CDPH samples and filtered USGS samples collected at wells sampled by both methods were used to create a raised threshold for unfiltered samples. The amount that the thresholds were raised by was calculated by taking the median value of the range of concentration differences between filtered and unfiltered samples and adding that value to the concentration threshold used for filtered samples. For example, if the differences in concentration between unfiltered and filtered Mn samples collected at the same wells within one year of each other yielded a range in differences of 0.05 to 0.17 mg/L with a median value of 0.07 mg/L, then the raised threshold for unfiltered Mn would be: filtered threshold (0.05 mg/L) + median difference (0.07 mg/L) = raised threshold (0.12 mg/L). Available data did not allow for classifications to be made based on SO_4^2 and CO_2 reducing conditions.

Table C1. Summary of groundwater age data and classification of samples into modern, mixed, and pre-modern age distributions, San Diego Groundwater Ambient Monitoring and Assessment (GAMA) study unit, California, May–July 2004.

[C, Degrees Celsius; TU, tritium units; –, no data]

GAMA_ID	Recharge temperature		Tritium		Terrigenic helium, percent of total helium	Modern carbon		Age classification
	(°C)	Error (°C)	(TU)	Error (TU)		(percent)	Counting error (percent)	
Grid Wells								
SDALLV-01	16.2	0.0	3.20	0.09	5.0	99	0.4	Mixed
SDALLV-02	15.9	0.0	4.08	0.18	0.5	99	0.4	Modern
SDALLV-03	17.0	0.0	0.00	0.07	90.2	59	0.3	Pre-Modern
SDALLV-04	19.2	0.1	2.28	0.11	7.6	–	–	Mixed
SDALLV-06	18.9	0.5	5.83	0.41	46.2	101	0.4	Mixed
SDALLV-07	16.0	0.0	4.31	0.18	7.4	–	–	Mixed
SDALLV-08	23.7	0.0	3.80	0.17	39.0	–	–	Mixed
SDALLV-09	20.4	0.2	0.01	0.03	81.6	22	0.2	Pre-Modern
SDALLV-10	15.7	0.0	5.81	0.22	18.8	–	–	Mixed
SDALLV-11	17.3	0.1	6.93	0.41	54.8	–	–	Mixed
SDALLV-12	18.3	0.0	2.65	0.12	83.6	–	–	Mixed
SDALLV-13	17.4	0.0	3.53	0.25	53.8	95	0.4	Mixed
SDALLV-14	15.9	0.0	2.16	0.10	28.2	–	–	Mixed
SDALLV-15	17.5	0.1	1.91	0.10	69.6	–	–	Mixed
SDALLV-16	21.0	0.8	4.92	0.41	0.0	–	–	Modern
SDHDRK-04	9.2	0.2	1.88	0.19	3.1	97	0.4	Modern
SDHDRK-05	18.3	0.0	3.32	0.31	63.7	102	0.4	Mixed
SDHDRK-06	12.7	1.3	0.41	0.05	96.4	76	0.4	Pre-Modern
SDHDRK-07	16.8	2.7	0.38	0.05	39.2	64	0.3	Pre-Modern
SDHDRK-08	14.0	0.3	1.92	0.11	72.3	–	–	Mixed
SDHDRK-09	14.7	0.6	1.93	0.10	30.2	–	–	Mixed
SDHDRK-10	15.1	0.7	2.82	0.13	10.4	–	–	Mixed
SDHDRK-11	17.4	0.0	1.00	0.19	84.4	–	–	Mixed
SDHDRK-12	14.1	15.1	0.50	0.19	1.2	–	–	Mixed
SDHDRK-13	15.5	1.5	2.82	0.31	21.4	–	–	Mixed
SDTEM-01	19.0	0.0	0.09	0.21	95.2	51	0.3	Pre-Modern
SDTEM-03	17.8	0.5	0.09	0.31	90.5	–	–	Pre-Modern
SDTEM-04	17.4	0.0	3.32	0.31	15.1	–	–	Mixed
SDTEM-05	20.9	0.0	0.00	0.15	58.3	84	0.4	Pre-Modern
SDTEM-06	16.3	0.0	1.50	0.09	22.0	96	0.4	Mixed
SDTEM-07	15.7	0.0	3.11	0.14	0.0	–	–	Modern
SDTEM-08	19.3	0.0	1.00	0.19	88.1	–	–	Mixed
SDTEM-09	13.8	0.2	0.03	0.04	99.1	–	–	Pre-Modern
SDTEM-10	17.1	0.1	6.43	0.41	0.0	91	0.4	Modern
SDTEM-11	17.6	0.0	0.04	0.04	82.6	–	–	Pre-Modern
SDTEM-12	16.9	0.0	0.91	0.07	94.0	98	0.4	Mixed
SDTEM-13	18.4	0.0	0.28	0.05	99.3	58	0.3	Pre-Modern
SDWARN-01	13.9	0.8	0.09	0.31	79.1	–	–	Pre-Modern
SDWARN-02	13.0	0.1	1.91	0.31	0.0	–	–	Modern
SDWARN-03	13.0	0.0	1.69	0.19	0.0	–	–	Modern

Table C1.   Summary of groundwater age data and classification of samples into modern, mixed, and pre-modern age distributions, San Diego Groundwater Ambient Monitoring and Assessment (GAMA) study unit, California, May–July 2004.—Continued

[C, Degrees Celsius; TU, tritium units; –, no data]

GAMA_ID	Recharge temperature		Tritium		Terrigenic helium, percent of total helium	Modern carbon		Age classification
	(°C)	Error (°C)	(TU)	Error (TU)		(percent)	Counting error (percent)	
SDWARN-04	13.1	0.0	0.19	0.19	94.3	78	0.4	Pre-Modern
SDWARN-05	12.8	0.0	0.09	0.19	23.7	85	0.4	Pre-Modern
SDWARN-06	16.0	0.4	0.06	0.05	5.8	81	0.4	Pre-Modern
SDWARN-08	11.9	0.1	0.21	0.05	98.8	–	–	Pre-Modern
SDWARN-09	12.0	0.0	0.04	0.05	98.7	–	–	Pre-Modern
Understanding Wells								
SDALLVU-1	17.1	0.5	1.84	0.09	99.2	–	–	Mixed
SDHDRKU-01	20.4	0.0	2.60	0.31	86.5	–	–	Mixed
SDHDRKU-02	16.1	0.0	0.32	0.06	0.0	–	–	Mixed
SDHDRKU-03	19.0	0.0	3.00	0.14	62.5	–	–	Mixed
SDTEMFP-01	17.6	0.1	0.28	0.04	93.5	74	0.4	Pre-Modern
SDTEMFP-02	14.2	0.0	1.10	0.19	95.1	70	0.3	Mixed
SDTEMFP-03	15.2	0.0	0.89	0.06	90.9	75	0.4	Pre-Modern
SDTEMFP-04	13.6	0.0	6.83	0.41	0.0	92	0.4	Modern
SDTEMFP-05	17.0	0.0	2.92	0.31	8.6	–	–	Mixed
SDTEMFP-06	14.4	0.0	1.25	0.19	95.1	–	–	Mixed
SDTEMFP-07	15.9	0.1	0.62	0.06	96.6	60	0.3	Pre-Modern

Table C2.   Concentration thresholds in milligrams per liter for dissolved redox indicator constituents used to infer redox conditions in groundwater, San Diego Groundwater Ambient Monitoring and Assessment (GAMA) study unit, California, May–July 2004.

[Modified from McMahon and Chapelle, 2008. Numbers in parenthesis are raised thresholds used for unfiltered samples. ≥, greater than or equal to; <, less than; –, data not available]

Redox condition	O_2	NO_3– as N	Mn	Fe
Oxic	≥ 0.5	≥ 0.5 (0.6)	< 0.05 (0.12)	< 0.1 (0.23)
Oxic	≥ 0.5	< 0.5 (0.6)	< 0.05 (0.12)	< 0.1 (0.23)
Anoxic	< 0.5	< 0.5 (0.6)	≥ 0.05 (0.12)	≥ 0.1 (0.23)
Anoxic	< 0.5	< 0.5 (0.6)	≥ 0.05 (0.12)	< 0.1 (0.23)
Anoxic	< 0.5	< 0.5 (0.6)	< 0.05 (0.12)	≥ 0.1 (0.23)
Anoxic	–	< 0.5 (0.6)	≥ 0.05 (0.12)	≥ 0.1 (0.23)
Anoxic	–	< 0.5 (0.6)	≥ 0.05 (0.12)	< 0.1 (0.23)
Anoxic	–	< 0.5 (0.6)	< 0.05 (0.12)	≥ 0.1 (0.23)

Appendix D. Comparison of CDPH and USGS-GAMA Data

Comparisons of CDPH and GAMA data were done to assess the validity integrating these datasets for the purpose of assessing water quality in the San Diego study unit. Because reporting levels for most organic constituents were substantially lower for data collected by the USGS than for data from the CDPH database (table 3), it generally was not possible to meaningfully compare concentrations of these constituent types in individual wells. However, because concentrations of inorganic constituents generally are detected at concentrations substantially above LRLs, a comparison of the two datasets was possible. Qualitative comparisons were done by plotting constituent concentrations on a one-to-one line graph and quantitative comparisons were done by calculating the relative percent difference (RPD) for each data pair (fig. D1). Only constituents with at least seven data pairs were examined.

Sample mass was sufficient for 15 inorganic constituents to allow comparisons to be made. Of these 15 constituents, the median RPD for 11 constituents was less than 15, for 3 constituents median RPD was either 23 or 24, and for 1 (iron) median RPD was 100 (fig. D1). Iron concentrations reported in the CDPH database were higher than the concentrations reported for samples collected by the USGS in five of seven replicate pairs. The differences in the sample collection procedures between the USGS and CDPH (filtered versus non-filtered) may be the reason for the higher concentrations observed in the CDPH samples than in the USGS samples. The results of this comparison show that inorganic data for most constituents from the CDPH database can be reasonably integrated with USGS data for the purposes of examining water quality in the San Diego study unit.

Major ion data for grid wells (USGS and CDPH data) were plotted on Piper diagrams (Piper, 1944) with CDPH major ion data to determine whether the grid wells represented the range of groundwater types that have historically been observed in the study unit. Piper diagrams show the relative abundance of major cations and anions (on a charge equivalent basis) as a percentage of the total ion content of the water. Piper diagrams are often used to define groundwater type (Hem, 1985).

The similarity of water types represented by CPDH data to water types represented by GAMA data indicate that the GAMA sampling design indeed did collect a representative sample of groundwater that is used for public supply in the San Diego study unit (fig. D2).

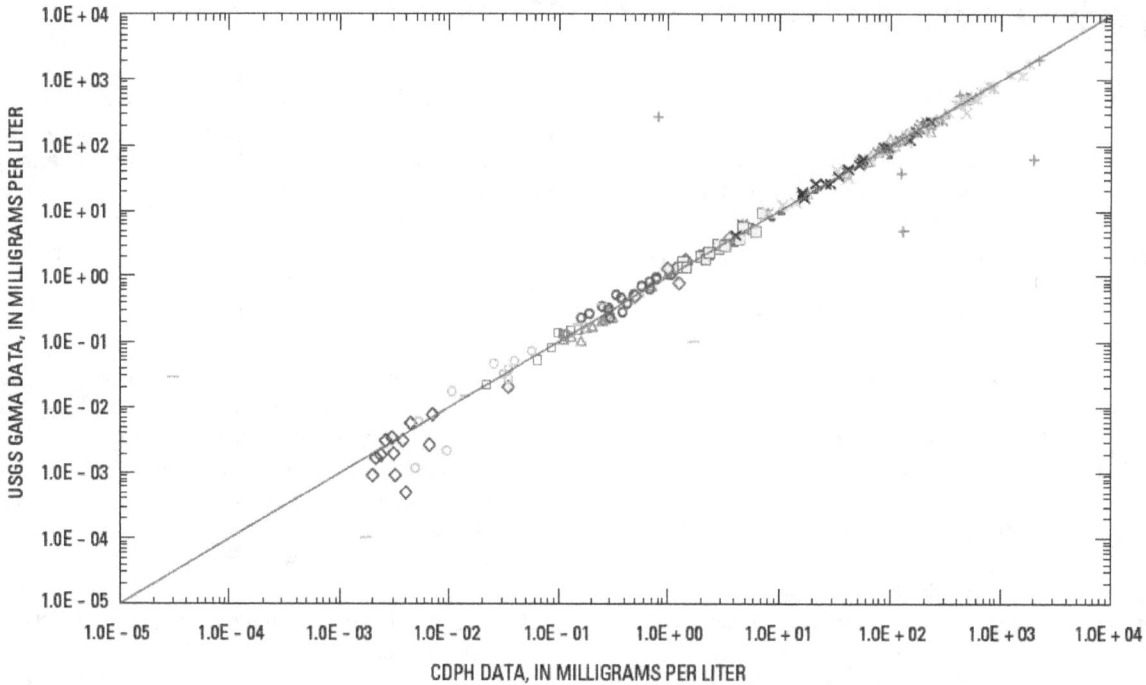

EXPLANATION

◇   **Arsenic** (median RPD = 23; n = 12)

□   **Barium** (median RPD = 11; n = 8)

△   **Boron** (median RPD = 13; n = 12)

✕   **Calcium** (median RPD = 4; n = 18)

✖   **Chloride** (median RPD = 5; n = 18)

○   **Fluoride** (median RPD = 14; n = 17)

+   **Iron** (median RPD = 100; n = 7)

-    **Magnesium** (median RPD = 7; n = 17)

—   **Manganese** (median RPD = 23; n = 23)

◇   **Nitrate** (median RPD = 10; n = 10)

□   **Potassium** (median RPD = 13; n = 17)

△   **Sodium** (median RPD = 5; n = 17)

✕   **Sulfate** (median RPD = 5; n = 18)

✖   **Total dissolved solids** (median RPD = 5; n = 18)

○   **Vanadium** (median RPD = 24; n = 10)

——   One-to-one line

Figure D1.   Paired inorganic concentrations from wells sampled by the Groundwater Ambient Monitoring and Assessment (GAMA) Program and the California Department of Public Health, San Diego Groundwater Ambient Monitoring and Assessment (GAMA) study unit, California, May–July 2004.

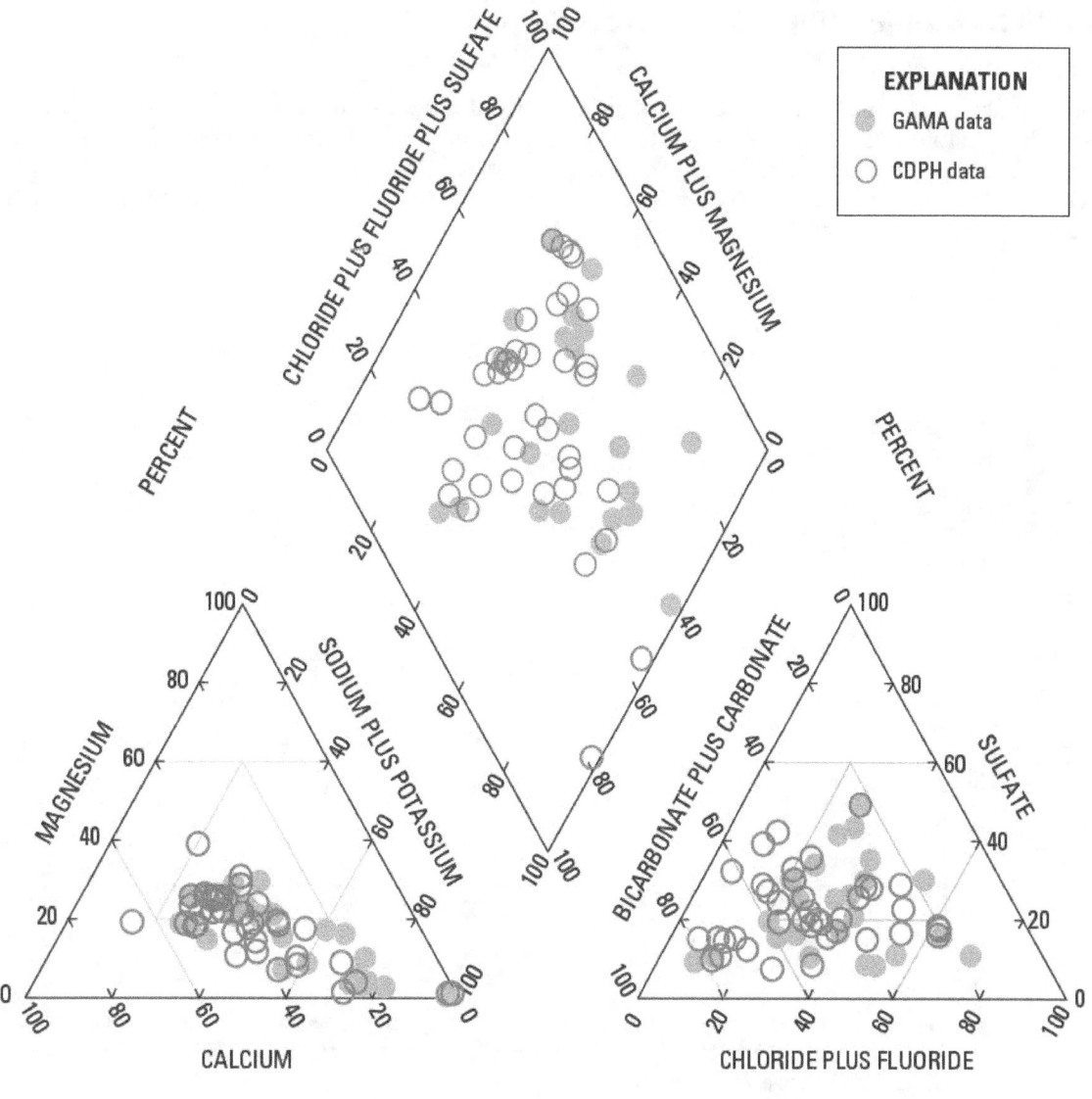

Figure D2. Piper diagram for grid wells and all wells in the California Department of Public Health database with a charge imbalance of less than 10 percent, San Diego Groundwater Ambient Monitoring and Assessment (GAMA) study unit, California, May–July 2004.

Appendix E. Calculating Total Dissolved Solids

Specific conductance, an electrical measure of TDS, was available in all 58 grid and understanding wells sampled by the USGS, whereas measured TDS data only were available for 24 of these wells. TDS values for the other 34 wells were calculated from specific conductance (SC) values using a linear regression equation (TDS = 0.628*SC –15.34) so that all grid wells would have TDS values. The predicted TDS values using the regression equation closely matched measured TDS values ($r^2 = 0.98$). TDS values from CDPH were combined with USGS measured and calculated TDS values.

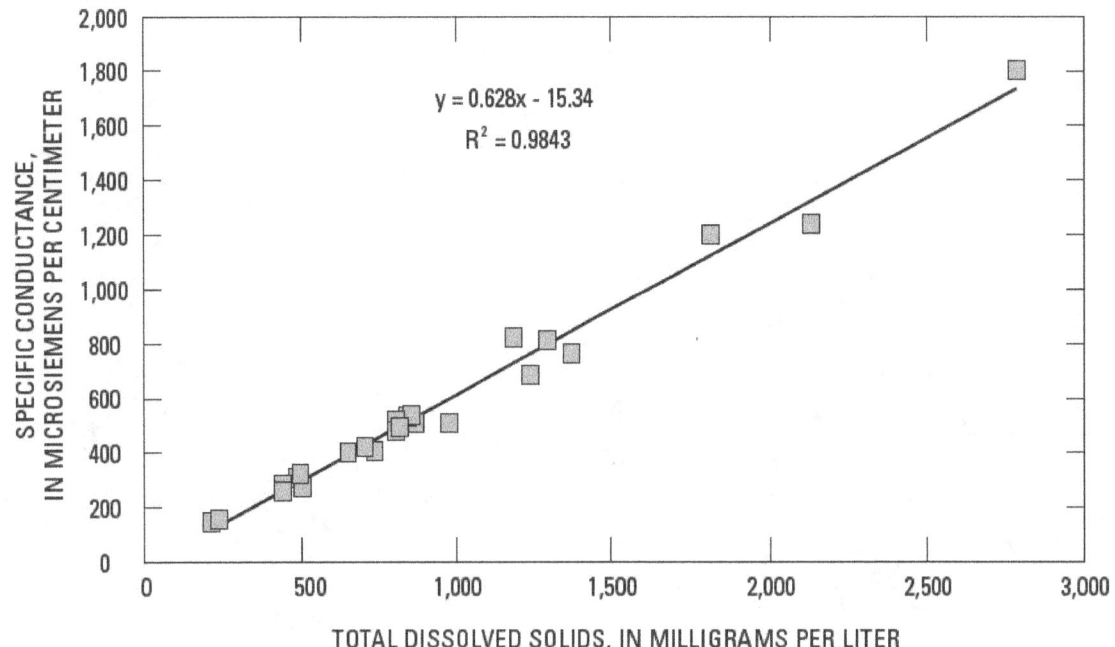

Figure E1.   Regression of total dissolved solids versus specific conductance for samples collected by the U.S. Geological Survey. San Diego Groundwater Ambient Monitoring and Assessment (GAMA) study unit, California, May–July 2004.